The Coral Reef

Life in the Sea

The Coral Reef

Pam Walker and
Elaine Wood

Facts On File, Inc.

The Coral Reef

Facts On File, Inc.
132 West 31st Street
New York NY 10001

Library of Congress Cataloging-in-Publication Data
Walker, Pam, 1958–
The coral reef / Pam Walker and Elaine Wood
p. cm.—(Life in the sea)
Includes bibliographical references and index.
ISBN 0-8160-5703-6 (hardcover)
1. Coral reef ecology—Juvenile literature. 2. Coral reefs and islands—Juvenile literature. I. Wood, Elaine, 1950– II. Title.
QH541.5.C7W35 2005
578.77'89—dc22   2004024225

Facts On File books are available at special discounts when purchased in bulk quantities for businesses, associations, institutions, or sales promotions. Please call our Special Sales Department in New York at
(212) 967-8800 or (800) 322-8755.

You can find Facts On File on the World Wide Web at
http://www.factsonfile.com

Text and cover design by Dorothy M. Preston
Illustrations by Dale Williams, Sholto Ainslie, and Dale Dyer

Printed in the United States of America

VB FOF 10 9 8 7 6 5 4 3 2 1

This book is printed on acid-free paper.

Contents

Preface

*L*ife first appeared on Earth in the oceans, about 3.5 billion years ago. Today these immense bodies of water still hold the greatest diversity of living things on the planet. The sheer size and wealth of the oceans are startling. They cover two-thirds of the Earth's surface and make up the largest habitat in this solar system. This immense underwater world is a fascinating realm that captures the imaginations of people everywhere.

Even though the sea is a powerful and immense system, people love it. Nationwide, more than half of the population lives near one of the coasts, and the popularity of the seashore as a home or place of recreation continues to grow. Increasing interest in the sea environment and the singular organisms it conceals is swelling the ranks of marine aquarium hobbyists, scuba divers, and deep-sea fishermen. In schools and universities across the United States, marine science is working its way into the science curriculum as one of the foundation sciences.

The purpose of this book is to foster the natural fascination that people feel for the ocean and its living things. As a part of the set entitled Life in the Sea, this book aims to give readers a glimpse of some of the wonders of life that are hidden beneath the waves and to raise awareness of the relationships that people around the world have with the ocean.

This book also presents an opportunity to consider the ways that humans affect the oceans. At no time in the past have world citizens been so poised to impact the future of the planet. Once considered an endless and resilient resource, the ocean is now being recognized as a fragile system in danger of overuse and neglect. As knowledge and understanding about the ocean's importance grow, citizens all over the world can participate in positively changing the ways that life on land interacts with life in the sea.

Acknowledgments

This opportunity to study and research ocean life has reminded both of us of our past love affairs with the sea. Like many families, ours took annual summer jaunts to the beach, where we got our earliest gulps of salt water and fingered our first sand dollars. As sea-loving children, both of us grew into young women who aspired to be marine biologists, dreaming of exciting careers spent nursing wounded seals, surveying the dark abyss, or discovering previously unknown species. After years of teaching school, these dreams gave way to the reality that we did not get to spend as much time in the oceans as we had hoped. But time and distance never diminished our love and respect for it.

We are thrilled to have the chance to use our own experiences and appreciation of the sea as platforms from which to develop these books on ocean life. Our thanks go to Frank K. Darmstadt, executive editor at Facts On File, for this enjoyable opportunity. He has guided us through the process with patience, which we greatly appreciate. Frank's skills are responsible for the book's tone and focus. Our appreciation also goes to Katy Barnhart for her copyediting expertise.

Special notes of appreciation go to several individuals whose expertise made this book possible. Audrey McGhee proofread and corrected pages at all times of the day or night. Diane Kit Moser, Ray Spangenburg, and Bobbi McCutcheon, successful and seasoned authors, mentored us on techniques for finding appropriate photographs. We appreciate the help of these generous and talented people.

Introduction

*C*oral reefs are opulent havens of life in the midst of relatively unproductive stretches of the ocean. Even though they are found in nutrient-poor waters, the rate of food production and animal growth in coral reefs is extremely high. Much of the success of reef life is due to the presence of one-celled algae living within the bodies of the tiny coral animals. These microbes help the corals by providing food and assisting in the construction of limestone skeletons. The coral skeletons themselves build structures that support some of the most diverse communities of life in the world.

The Coral Reef is one of six books in Facts On File's Life in the Sea series, which examines the physical features and biology of different regions of the ocean. *The Coral Reef* focuses on the organisms that make up these specific communities. Chapter 1 reviews the history of reef structures across geologic time, paying particular attention to the key roles of cyanobacteria and stromatolites. The geologic forces that created coral reefs and the factors involved in reef evolution are included in this chapter. Distinct zones of the coral reef reveal physical characteristics that are tied to the types of life those zones support. Life on every reef is dependent on the geological qualities, as well as physical and chemical factors such as temperature, salinity, available light, dissolved gases, and nutrients contained in the water column.

The living things that make their homes in, on, and around the reef create an ecosystem whose biodiversity rivals that of the tropical rain forests. Chapter 2 explores how these organisms are supported by the producers in the system, the lifeforms that contain chlorophyll and other photosynthetic pigments. Unlike terrestrial ecosystems that are supported by large plants, the primary producers in the coral reef are microscopic, one-celled protists, cyanobacteria, and a few species of macroalgae. In addition, decaying organic matter forms the foundation for a rich community of detritivores. Both producers and detritivores serve as food for the small organisms of the reef and form the basis of numerous complex food chains.

Among the most numerous of reef consumers are the invertebrates, small animals that lack backbones, the topics of chapters 3 and 4. The coral animal itself is an invertebrate that lives in a calcium carbonate skeleton of its own making. Among the corals are the mollusks, organisms that have a muscular foot that is used for locomotion, a sheet of tissue over their organs called a mantle, and in many cases, an external shell. They include clams, mussels, snails, and nudibranchs that hide in the reef bottom, as well as octopuses, squid, and cuttlefish. Arthropods are also numerous, and their populations include shrimps and lobsters. The reef floor is dotted with spiny-skinned animals, the echinoderms. Distinguished by their star-shaped bodies, echinoderms inch across the reef on tube feet, consuming mollusks as they travel.

The largest reef consumers are vertebrates: fish, reptiles, birds, and mammals. All of these animals are highly mobile, some living in or near the reef year round. Others just winter at the reef when temperatures are too cool outside the tropical waters. Fish, the topic of chapter 5, are probably the most visible vertebrates, and those that live near the reef show a variety of structural adaptations for life in this unique habitat. Because of their large populations, competition for food among fish is intense. For this reason, adaptations for feeding and reproducing are varied and often extreme. Typical adaptations include the parrot fish's beaklike mouth, a perfect instrument for biting off alga and bits of coral and the moray eel's long, finless body, highly adapted for swimming through small spaces. Many fish flash bright colors, some of which are intended to warn away predators, others designed to attract mates.

Chapter 6 discusses reptiles and birds, animals that are as finely modified for reef life as the resident fish. Sea turtles and sea snakes are reptiles that are quick and graceful swimmers. Seabirds lay their eggs on the beaches of coral reef islands and find their prey in the nearby waters. The other top predators in the reef ecosystems are humpback whales, minke whales, and spinner dolphins. With plenty of prey to feed on and warm waters in which to swim, they occupy the top position in many richly populated food chains.

Chapter 7 underscores the fragile state of coral reef ecosystems. Losses of reefs due to human activities have prompted national and international groups to monitor these regions and safeguard their inhabitants. In marine sanctuaries, where reefs receive protection, communities of life are thriving and growing.

As in every ecosystem, reef producers and consumers play roles in the ongoing stories of life and death. In all probability, every animal born on the coral reef will be consumed by another animal. The unconscious goal of each animal is to eat, mature, and reproduce during its time on Earth. The strategies that living things have found to ensure their survival are testaments to the ability of life to adapt and continue.

Physical Aspects
Structure and Science of the Coral Reef

*E*ach year divers, fishermen and -women, scientists, and sightseers visit coral reefs. These brightly colored marine communities are found off the coasts of more than 100 countries, including the United States, Australia, India, China, Japan, Mexico, and Belize. At first glance, the reefs appear to be magnificent underwater structures built from stone. Closer inspection reveals that these aquatic complexes are actually composed of millions of living organisms resting atop the skeletons of their ancestors.

The living and growing parts of the reef form only a thin veneer on top of the remains of dead corals, algae, mollusks, and sponges. As organisms die, they leave behind their skeletons, expanding the base on which the next generation builds. Over thousands of years, coral reefs grow to gigantic sizes, reaching lengths of several miles.

Although visits to coral reefs reveal colossal structures and abundant life, these systems are rare, occurring in less than 0.4 percent of the ocean's waters. Their scarcity is due to their requirements for precise physical conditions. Reefs develop and thrive in seawater within a narrow range of temperatures. Coral animals require some nutrients but are intolerant of extremely high levels. The water of reefs must be energetic enough to dissolve and incorporate oxygen, and it must be shallow enough to be penetrated by light. This unique set of conditions is most likely to occur in locations near the equator.

People are interested in coral reefs for a variety of reasons. Many gain their living from these aquatic gardens, harvesting their bounty or marketing their beauty. Some coastal communities are protected from the brunt of the ocean's forces by the barrier provided by the reef's physical structure. Leaders in the fight against disease are exploring the reef's collection of unique chemicals, looking for those with potential as medications. As reefs gain attention, citizens of the world are becoming increasingly aware of the uniqueness and fragility of these ecosystems. More and more, coral reefs are being recognized as wild places whose existence may be endangered by human activities. The key to their survival may hinge on humankind's ability to understand them better.

Covering only about 108,000 square miles (about 280,000 sq km) in total, reefs make up a relatively small part of the ocean; however, they are remarkably important ecosystems, supporting more than 25 percent of all known marine species. Coral reefs serve as homes, nurseries, feeding grounds, and gathering places for thousands of kinds of living things, such as the pyramid bluefish in Figure 1.1. The great variety of organisms found among the coral reefs makes them the most biodiverse marine ecosystems on the planet. For that reason, some scientists refer to them as "the tropical rain forests of the ocean" because, like rain forests, reefs support great biodiversity.

Fig. 1.1   The pyramid bluefish is one of hundreds of brightly colored species that live on coral reefs. (Courtesy Getty Images)

Biodiversity

Biodiversity, or biological diversity, refers to the variety of living things in an area. Diversity is higher in complex environments than in simple ones. Complex physical environments have a lot to offer organisms in the way of food and housing. Estuaries, shorelines, and coral reefs are extremely complex marine environments, and each of them provides a wide assortment of nutritional resources for living things.

There are thousands of habitats in estuaries, coastal systems where fresh and salt water meet and mix. The bottom of the estuary provides homes for different kinds of organisms. Some spend their entire lives on the surface of the sediment, many burrow just under the surface, and others dig deep into the sediment. Organisms also select locations that accommodate their abilities to tolerate salt, so those that are adapted to high salinity are on the seaward side while the freshwater-dependent ones are on the river side. In between the two extremes, organisms live in zones that meet the salinity requirements for their bodies.

Diversity is an important aspect of a healthy ecosystem. In an ecosystem where all living things are exactly the same, one big change in the environment could cause widespread destruction. This might be best understood in a familiar ecosystem, like a forest. If only one kind of tree is growing in the forest, a virus that damages that type of plant could wipe out the entire forest. If the forest contains 20 different kinds of trees, it is unlikely that one disease agent could destroy the entire plant community. A high degree of biodiversity gives an ecosystem an edge, ensuring that it can continue to exist and function regardless of changes around it.

Despite their impressive biological and physical diversity, coral reefs must remain in balance to flourish. The equilibrium of nonliving factors such as sunlight, nutrients, and temperature with living factors such as population size and food supply constantly adjusts and fine-tunes itself. As in any ecosystem, each part of the reef community is dependent on its other parts. If one component of the reef is disturbed, the entire community has to adjust.

For the observer, an opportunity to view reef organisms in their environment is like attending a living museum in natural history. Some reefs are homes to types of organisms that have been in existence for thousands of years. These life-forms boast genealogies longer than any organisms in land-based ecosystems. Some of the present-day coral reefs were thriving when the land adjoining them was first populated with humans. Reefs have played an important cultural role in developing nations and are part of the history of the sea and lands they border.

Carbon Dioxide Grabbers

Coral reefs help keep the Earth's biosphere, the part of the planet where living things are found, in balance. One of the coral reef's important functions is in maintaining normal levels of carbon dioxide in the atmosphere. At the point where the atmosphere meets the sea, carbon dioxide and other gases from the air dissolve in ocean water. In places where coral reefs exist, much of this dissolved carbon dioxide is removed from the water by coral organisms. The organisms then use the gas to build calcium carbonate, or limestone, skeletons. As the skeleton-building proceeds, levels of the dissolved gas in ocean water decrease, permitting

Greenhouse Gases

Carbon dioxide is one of several so-called greenhouse gases that form an invisible layer around the Earth. As shown in Figure 1.2, greenhouse gases trap the Sun's heat near the Earth's surface, very much like the windows in a greenhouse hold in heat from the Sun. The greenhouse gases are one of the reasons that temperatures on Earth's surface are warm enough to support life. If they did not exist in the atmosphere, most of the Sun's radiant energy would bounce off the Earth's surface and return to space.

The layer of greenhouse gases is changing, however, and this change has many scientists worried. By burning fossil fuels in homes, cars, and industries, people all over the world are constantly adding carbon dioxide to the air, widening the belt of greenhouse gases. Many environmentalists fear that the rising levels of carbon dioxide in the air are warming the Earth's surface abnormally, a phenomenon known as global warming.

Research indicates that some warming has already taken place in the air and in the ocean. The effects of this warming include less snow cover each winter, a retreat of mountain glaciers, and changes in global weather patterns. Experts fear that continued warming could damage the balance of life on Earth. Some predict far-reaching results, including changes in climates, melting of glacial ice, and damage to the coral reefs.

Fig. 1.2   Carbon dioxide is one of the greenhouse gases in the atmosphere that traps heat close to the surface of the Earth.

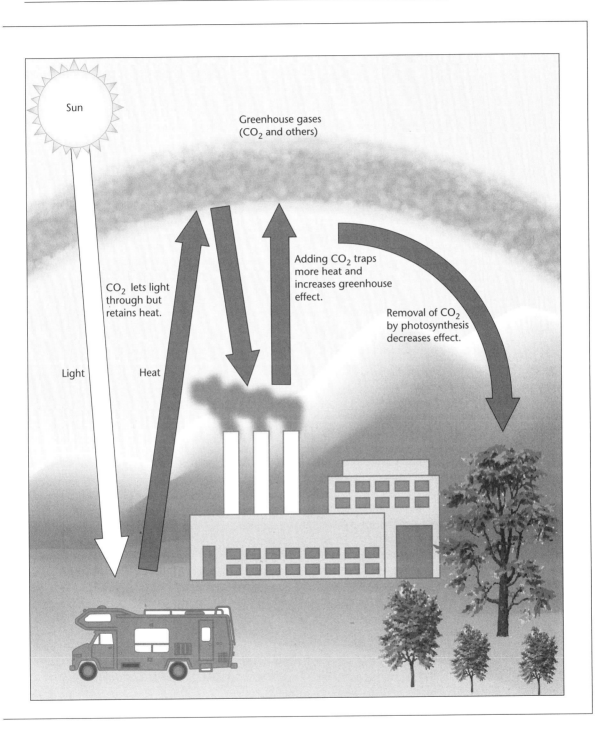

more carbon dioxide to enter the water from the atmosphere. For this reason, reefs act as carbon "sinks."

Coral reefs are usually found near coastlines. Because of their positions in relation to landmasses, some of them form natural, protective walls for coasts. The walls act as fortresses, diminishing the destructive forces of the waves as they pound the shore during storms or times of high tides. These reef walls also help prevent erosion, damage to coastal sea life, loss of property, and even loss of human life. Without the coral reefs, the homes and businesses of millions of people would be exposed to the full fury of the sea. About one-sixth of the world's shores are protected by reefs. Some of these areas, such as the coasts in Asia, support the densest populations of humans in the world.

Coral reefs also contribute to beach formation. Natural forces break off pieces of the reef and grind them into grains of sand. As wind and water strike the reef, they chip away at the skeletal structures of reef animals, eroding them into small pieces. Predators also loosen reef material by nibbling on it to get at choice foods. Even some of the plants and animals that grow on reefs erode them. Once dislodged, small particles of reef are tossed and crushed by waves until they form fine particles of sand. Beaches created primarily by erosion of coral are brilliantly white. Barbados, an island in the West Indies, is one of hundreds of islands built on coral and famous for its prized white beaches.

Origins of Coral Reefs

A visitor to a coral reef millions of years ago would have witnessed a seascape that is quite different from the one that exists today. Over time, both the appearance and composition of reefs have changed dramatically. Reefs have been subjected to countless alterations over their history. Ice ages, mass extinctions, shifting of landmasses on continental plates, and fluctuating sea levels are just a few of the global events that reefs have endured.

Geologic records document the existence of reefs 2 billion years ago, in a period of time referred to as the Precambrian

Geologic Time

The Earth is about 4.5 billion years old. Fossil evidence suggests that the first living things were simple cells that appeared about 3.5 billion years ago. The time line in Figure 1.3 shows that the period of time from the beginning of Earth to 700 million years ago, the largest part of the Earth's past, is known as the Precambrian era. The Paleozoic era began about 570 million years ago and lasted until 280 million years ago. Fish, insects, amphibians, and reptiles were some of the major groups of animals that developed in this period. Both terrestrial and aquatic plants also formed in this time span. The Mesozoic era extended from 250 million years ago until 135 million years ago. A period dominated by reptiles, the Mesozoic is known as the age of the dinosaur. Late in the era, mammals and birds developed. The most recent period, the Cenozoic era, began 65 million years ago and extends to the present. During this time, birds and mammals flourished. Humans made their appearance late in the era, about 3 million years ago.

To visualize the amount of time that has passed since the first coral reef appeared on Earth 2 billion years ago, one can compare time to a human's walking stride. For example, a person's stride, a distance of about 3 feet (0.9 m), could represent a period of 50 years. In such an analogy, walking two steps back would take one back a century in time. The distance of 40 steps would represent the time that has passed since the birth of Jesus (the beginning of the Christian era [C.E.]), and 200 strides would bring one to human's prehistoric period. Yet, to reach the time when reefs were first formed on Earth, one must walk a distance equal to the Earth's circumference at the equator (24,902 miles [40,076 km], or 43,827,520 strides)!

Era	Age (millions of years)	First life-forms
Cenozoic	0.01	
	3	Humans
	11	Mastodons
	26	Saber-toothed tigers
	54	Whales
	65	Horses, alligators
Mesozoic	190	Birds
	210	Mammals
	230	Dinosaurs
Paleozoic	280	Reptiles
	345	Amphibians
	400	Sharks
	435	Land plants
	500	Fish
	570	Sea plants
	650	Shelled animals
Precambrian	700	Invertebrates
	3500	Earliest life
	4600	

Fig. 1.3   The geologic time scale shows significant events in the development of life on Earth.

era. The architects of the ancient reefs were not coral but simple microbes called cyanobacteria. Then, as now, cyanobacteria were algaelike organisms that formed long, mucus-producing filaments. Their sticky filaments trapped and held debris and grains of sand. Individual algae, with their ensnared soil particles, stuck to one another, forming tall, gray towers, or stromatolites, that rose several meters upward from the seafloor. From 2 billion years ago to 500 million years ago, a period of 1.5 billion years, stromatolites flourished near coastlines.

About 600 million years ago, cyanobacteria were joined in their reef-building activities by archaeocyathids, spongelike animals with stony textures. The word *archaeocyathid* means "ancient cup" and aptly describes the appearance of these simple animals. The union of blue-green algae and these primitive animals yielded reefs of great durability. The partnership between the two lasted for the next 60 million years until communities were severely damaged by the first of many mass extinctions that have occurred in Earth's history. Eventually, cyanobacteria and their stromatolite structures alone made a comeback, and the more primitive-style reefs returned.

Around 480 million years ago cyanobacteria teamed up with an animal more complex than the simple archaeocyathids. The new partners were bryozoans, animals with a mosslike appearance. Soon afterward, stony sponges, red algae, and the first of the true corals developed. All four types of organisms were capable of building a limestone covering over their bodies, a feature that protected them from the destructive action of the ocean waves. When these creatures died, their lacy, branching shapes added new dimensions to the reef structure. The association of cyanobacteria's stromatolites with this new team of skeleton-making organisms lasted for 130 million years.

Around 350 million years ago, thousands of living things, including many species of corals, bryozoans, red algae, and sponges, were wiped out by a second mass extinction. Again, only the hardy cyanobacteria and their stromatolites survived. For the next 13 million years the cyanobacteria existed alone, once again building their drab, gray towers. Eventually, several species of calcium carbonate–secreting green algae, stony sponges, and bryozoans joined them.

This latest wave of reef building continued for 115 million years, until another mass extinction struck 225 million years ago, claiming over half of the planet's plants and animals. During this time several coral species suffered destruction and the stromatolites were reduced vastly in number, causing the reef population to once again disappear.

Reef builders were not able to start over for another 10 million years. When they did appear again, new families of coral, the ancestors of today's coral populations, developed. For 130 million years, reefs expanded their locations, spreading from a few scattered sites to areas all around the world. At the same time, reef inhabitants changed. New varieties of sponges and mollusks moved in, and the role of cyanobacteria and their stromatolites as primary reef builders declined. Red algae, which had teamed with several new species of coral, acted as the major architects of the reefs of this time period.

About 65 million years ago a final mass extinction annihilated life-forms around the Earth. During this period, one-third of animal species, including the dinosaurs and many species of coral and other reef-building organisms, were lost. Ten million years passed before the reefs reappeared. The coral reefs that made a comeback grew vigorously and have endured till this day.

Physical Characteristics of Coral Reefs

Coral animals can be found in several parts of the ocean, but the reef-building types only live in places that meet a narrow range of environmental conditions. Reef-building corals have very specific habitat requirements. They are finicky about the amount of salt in the water, water temperature and depth, movement of currents, and available nutrients.

Salinity refers to the amount of dissolved minerals, or salts, in ocean water. The average salinity of ocean water is 35 parts per thousand, which can be written as 35‰. The symbol ‰ is similar to percent but refers to parts per thousand instead of parts per hundred. Salinity is low in areas where freshwater flows into the ocean, such as near the mouths of rivers. Salinity is high in places where water evaporates from slow-moving or stagnant pools of salt water.

Reef-building corals favor waters where the salinity is about 34 parts per thousand by weight, a little lower than average sea salinity. Coral reefs do not exist in places where freshwater runs into the ocean and drastically reduces the salinity. That is why there are no coral reefs in the part of the Atlantic Ocean where the Amazon River meets the sea, even though other physical factors of the region are ideal.

Although some species of coral can be found in deep, cold ocean waters, stony coral, the type that forms hard skeletons, primarily exist in warm ocean waters. Some reef-building coral species are hardier than others, but water temperatures between 68°F (20°C) and 96.8°F (36°C) are suitable for most, with 75.2°F (24°C) being the ideal. For this reason coral reefs

Chemical and Physical Characteristics of Water

Water is one of the most wide-spread materials on this planet. Water fills the oceans, sculpts the land, and is a primary component in all living things. For all of its commonness, water is a very unusual molecule whose unique qualities are due to its physical structure.

Water is a compound made up of three atoms: two hydrogen atoms and one oxygen atom. The way these three atoms bond causes one end of the resulting molecule to have a slightly negative charge, and the other end a slightly positive charge. For this reason water is described as a polar molecule.

The positive end of one water molecule is attracted to the negative end of another water molecule. When two oppositely charged ends of water molecules get close enough to each other, a bond forms between them. This kind of bond is a hydrogen bond. Every water molecule can form hydrogen bonds with other water molecules. Even though hydrogen bonds are weaker than the bonds that hold together the atoms within a water molecule, they are strong enough to affect the nature of water and give this unusual liquid some unique characteristics.

Water is the only substance on Earth that exists in all three states of matter: solid, liquid, and gas. Because hydrogen bonds are relatively strong, a lot of energy is needed to separate water molecules from one another. That is why water can absorb more heat than any other material before

are predominately scattered throughout the tropical and subtropical western Atlantic and Indo-Pacific Oceans between the tropics of Cancer and of Capricorn. These are the areas of the world that experience only small changes in weather between seasons. In the tropical Pacific Ocean, the reefs are widely distributed, but in the western Atlantic Ocean they are confined to the Florida Keys, Bermuda, the Bahamas, the Gulf of Mexico, and areas in the Caribbean Sea.

The number of different coral species that compose a reef is dependent on the ocean in which the reef is located. The Indo-Pacific coral reefs are rich in species diversity, boasting more than 500 different coral species, while the Atlantic Ocean reefs are made up of between 60 and 70 coral species. Scientists are not sure why there is such a difference in species diversity in the two locations, but they suspect that

its temperature increases and before it changes from one state to another.

Since water molecules stick to one another, liquid water has a lot of surface tension. Surface tension is a measure of how easy or difficult it is to break the surface of a liquid. These hydrogen bonds give water's surface a weak, membrane-like quality that affects the way water forms waves and currents. The surface tension of water also impacts the organisms that live in the water column, water below the surface, as well as those on its surface.

Atmospheric gases, such as oxygen and carbon dioxide, are capable of dissolving in water, but not all gases dissolve with the same ease. Carbon dioxide dissolves more easily than oxygen, and there is always plenty of carbon dioxide in seawater. On the other hand, water holds only $\frac{1}{100}$ the volume of oxygen found in the atmosphere. Low oxygen levels in water can limit the number and types of organisms that live there. The concentration of dissolved gases is affected by temperature. Gases dissolve more easily in cold water than in warm, so cold water is richer in oxygen and carbon dioxide than warm water. Gases are also more likely to dissolve in shallow water than deep. In shallow water, oxygen gas from the atmosphere is mixed with water by winds and waves. In addition, plants, which produce oxygen gas in the process of photosynthesis, are found in shallow water.

the most recent ice age was more damaging to the Atlantic Ocean reefs than to others.

Temperature affects corals in several ways. The coral animals constantly convert dissolved carbon dioxide and calcium into calcium carbonate, a compound that forms their skeletons. In warm water, calcium carbonate reaches saturation levels very quickly. At saturation, a dissolved compound precipitates, changing from a dissolved form to a solid one. The ability to convert dissolved calcium carbonate to the solid form helps corals create plenty of skeletal material. On the other extreme, if water temperatures get too high, the consequences are disastrous. Coral are unable to create any calcium carbonate to build or repair skeletons. In addition, corals cannot reproduce in water that is too warm.

Because reef-building corals form important relationships with microscopic green organisms, they grow best if they receive plenty of sunlight. Sunlight does not penetrate water deeper than 150 feet (about 46 m), so corals cannot grow below that depth.

In addition, coral reefs are very sensitive to the amount of dissolved nutrients in the water. Coral animals thrive in nutrient-poor conditions, because high levels of nutrients can stimulate the growth of tiny marine plants. Overgrowth of these water plants, a phenomenon called algal bloom, can make the water dark and murky, preventing corals' resident algae from receiving enough light. Nutrient-poor, or oligotropic, waters, which are characteristically blue in color, are typical of coral reefs.

The distribution and growth of reefs in the ocean is also influenced by the flow of ocean currents. Clear, moving water is extremely important to the survival of reef-building coral. Moving water carries food, nutrients, and oxygen to the living coral animals. Reefs are rarely found where there are large amounts of suspended matter because debris, silt, or other particulates can smother the fragile coral animals.

In the Zone

No two coral reefs are exactly alike. Each one is a dynamic and ever-changing structure. Despite their differences, most coral reefs display several distinct zones that are created by environmental conditions such as wave and current strength, suspended sediment content, temperature, and depth of the water. Zones vary somewhat, depending on ocean location and type of reef, but most reefs have four typical zones: the reef flat, reef crest, buttress, and seaward slope.

The part of the reef that is closest to the shore is called the reef flat, or back reef. In this area, living things are protected from the full force of the breaking waves; however, water on the reef flat is relatively shallow, ranging in depth from a few centimeters to a couple of meters. Shallow-water inhabitants are exposed to wide variations in temperature and salinity. They must also deal with changing water levels and occasional periods when the low tide leaves them stranded without water. These factors limit the types of organisms that survive in the reef flat.

Moving from the shore toward the ocean, the second zone is the reef crest, also called the algal ridge. This is the highest point of the reef, and it is exposed to the full impact of waves that rush from the ocean toward the shore. During times of low tide, this area is fully exposed to the penetrating rays of the Sun. As in the reef flat, only a limited number of organisms can survive in this zone.

The third zone, traveling seaward, is the buttress. This area begins at the point where low tide waters cover the reef and continues out to a depth of about 65.6 feet (20 m). Jagged extensions of the buttress zone jut from the reef out into the ocean. The undulating shape of the buttress zone diverts water striking the reef into many direction. By spreading the impact of the waves, the reef buttress helps the structure withstand their full power and impact. Channels in the buttress drain debris and sediment out to sea. With plenty of sunlight and oxygen present, huge reef-building coral and algal colonies develop in the upper part of the buttress zone. The corals that grow on top of the buttress tend to develop short, thick branches, while those further under the water look like small shelves or branched plants.

The final zone, the seaward slope, begins where the buttress zone ends, just below 65.6 feet in depth. The upper section of

Coral Reefs

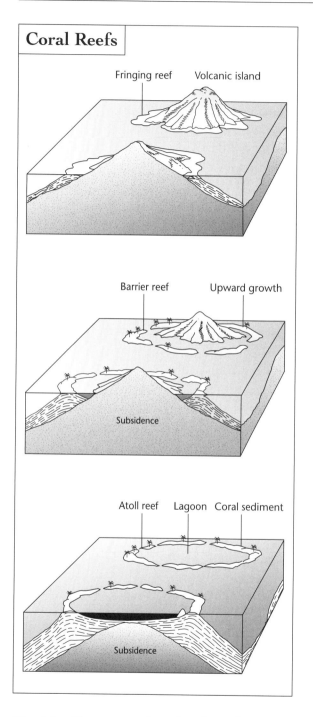

Fig. 1.4　Three types of coral reefs are fringing reefs, barrier reefs, and atoll reefs.

this area, which receives the most sunlight, holds many different species of coral. Below 131.2 feet (40 m) deep, fewer are found because sediment builds up in the water, blocking the light. This deepwater region supports a lot of sponges and non-reef-building corals.

All of the zones of coral reefs support complex groups of living things, including more than 3,000 different kinds of animals. Competition among these organisms for available food and space is intense. Some species may overgrow and squeeze out others in an effort to utilize nutrients and light. Many other animals share the space effectively by limiting the times they forage for food; for example, some only come out at night, and others are active in the daytime.

Types of Coral Reefs

Depending on where they are located and how they are formed, shallow-water tropical reefs can be classified into one of three major groups: fringing reefs, barrier reefs, and atoll reefs. Figure 1.4 illustrates the structure of each reef type. Fringing reefs, which form along a coastline, are the most common type. These develop at the margin of a landmass where conditions are suitable for coral growth. They are normally located only in shallow waters and border the coast very closely with only a narrow stretch of water separating the reef from the shore. Because sediment washes from the land out to the sea, most fringing reefs have very little coral growing on the shore side. However, the ocean side, which is not exposed to as much sediment, is home to large populations of live coral. Fringing reefs are common in the Caribbean and around the Hawaiian Islands.

Like fringing reefs, barrier reefs run parallel to the shore-line, but they are located further out in the ocean. A barrier reef is separated from the shoreline by a lagoon, a deep, open body of water with a sandy bottom. Lagoons are home to many forms of life. The shallow sections contain large under-water fields of grass. The root systems of these plants help to trap sand, further adding to the base of the lagoon. The barri-er reefs are so named because they form a barrier between the lagoon and the ocean. The largest reef in the world, the Great Barrier Reef, is located off the eastern coast of Australia. The Great Barrier Reef is more than 500,000 years old.

The Great Barrier Reef

The largest reef in the world is the Great Barrier Reef located off the coast of Australia. Bigger than the entire country of Italy, this reef system measures 1,249.1 miles (2,011 km) in length and 44.7 miles (72 km) across at its widest point. The reef is not a continuous struc-ture but is made of more than 2,800 indi-vidual reefs. More than 400 types of coral, 1,500 species of fish, 4,000 types of mollusks, and 400 kinds of sponges make their homes in the Great Barrier Reef. Other animals there include anemones, worms, crustaceans, and echinoderms.

The reef supports sea grass beds that are feeding grounds for the dugong, an endangered mammal, as well as for the endangered green and loggerhead tur-tles. The reef is also used by humpback whales that travel from the Antarctic to give birth in its warm waters.

The Great Barrier Reef is the best-known marine protected area in the world. Because it is a living classroom of natural history and science, the reef was declared a marine park in 1975 to pre-serve its condition while providing rea-sonable use.

The third classification, the atoll reef, is made of circular coral structures. These formations grow on top of volcanoes that lie below the ocean surface. Like barrier reefs, atolls surround central lagoons. These coral reefs are commonly found in the Indo-Pacific regions with the largest atoll being Kwajalein, which surrounds a 60-mile (97-km) wide lagoon.

Evolution of a Coral Reef

Scientists have studied the structures of coral reefs for decades, trying to determine how they were formed. The theory that most present-day scientists accept was among the earliest proposed. Naturalist Charles Darwin first presented his ideas on reef evolution in the 1830s.

Darwin believed that coral reefs changed over long periods of time, evolving from fringing reefs to barriers and finally to atolls. He explained that the process began when the eruption of an active volcano in the ocean created a small island of lava. After the volcano became inactive, it cooled, leaving part of its surface (the island of lava) jutting above sea level. At first this tip of the volcanic mountain lacked life. Ocean waters carried immature coral animals to the mountain island's rough, rocky shores. These young corals attached to the volcano in the shallow waters and grew into adults with hard skeletons. As the corals grew and reproduced, they spread around the entire volcanic island, eventually creating a substantial fringing reef.

Meanwhile, the volcanic mountain began to sink gradually into the sea, taking part of the coral reef to deeper water. As the reef continued to develop, it grew in the direction of the water's surface and the sunlight. At some point, the volcanic base sank to such depths that some of the coral animals on the island side of the reef could no longer receive enough light, and they died. As landward portions of reef disappeared, the body of water between the reef and the island increased in size, looking very much like a barrier reef by today's standards.

As time passed, the volcanic mountain continued to sink until even its tip was completely submerged below the surface of the water. However, coral kept on growing on top of the submerged reef. Eventually, all that could be seen above water was a ring of coral surrounding a lagoon. As the years passed, sand was trapped by the reef, creating beaches. This partial ring of coral became an inhabitable island, an atoll.

The atoll must constantly deal with the destructive forces that threaten it. Seawater and rain slowly dissolve its limestone base. Animals searching for food nibble at the reef to get the algae embedded in it, weakening its structure. Strong waves break apart pieces of limestone and wash silt, sand, and coral debris into the lagoon.

However, nature's forces do not just erode the reefs; they also sculpt and remodel them. Wind and waves grind up the coral debris and sediment to form sand that finds its way to beaches on coral islands. The sandy shore provides a home for seeds that make their way to the beach by way of the wind and birds. Eventually plants and trees begin to grow. Only very hardy trees can gain a foothold on the side of the lagoon that receives the brunt of strong winds and high waves. A greater variety of plants, such as coconut and breadfruit trees, are found in the more protected interior portions of the lagoon.

The growth of trees on the new island helps to further develop its shape, and hold sand on its shores. A natural fertilizer, bird guano (excrement) enriches the soil as more and more bird species find their way to the new island. Eventually, a multitude of animal life inhabits the lagoon. Female turtles lay their eggs on the shore, bats feed on the fruit of trees, and small lizards dart across the debris-covered areas of the island. The atoll becomes an island that serves as the home to a large number of plants and animals.

Deep Water Reefs

Although the best known reefs are those in warm, tropical waters, coral reefs exist in other locations. Deepwater coral reefs can be found near landmasses around the globe in waters from 656.17 feet (200 m) to 4,921.26 feet (1,500 m) in depth. Deepwater reefs are similar in many ways to those in shallow waters. The most obvious differences in the two environments are temperature and available light.

A rocky or firm surface provides deepwater coral animals a point of attachment. For this reason, most reefs in deep water are located on underwater mounds, ridges, slopes, and mountains. Strong, fast-moving currents are almost always associated with these communities because they continuously supply water that is laden with oxygen and particles of food. Strong currents also help disperse the reproductive cells of corals and keep their surfaces free of sediments.

Scientists have been aware of deepwater reefs for more than 200 years, but gathering information on them has been a challenge. These habitats are widely scattered throughout the ocean and located at depths that make them hard to study. A reef associated with the Dry Tortugas, a cluster of islands near Key West, was first sighted in 1999 by a team of researchers from the University of South Florida. Located on an underwater barrier island called Pulley Ridge off the southwest coast of Florida, this reef is the deepest in U.S. waters. Little was known about Pulley Ridge until 2004 when further studies showed it to be a thriving deepwater community. Unlike other deepwater reefs, Pulley Ridge is the only one known to be dependent on light filtering from the surface. Because the light at this depth is extremely low, corals, sponges, and algae assume flattened shapes to maximize their surface area. In shallow-water systems, corals structures are taller and thinner.

Conclusion

Each coral reef is a unique and highly productive ecosystem. A reef can support thousands of different species from almost every known group of living things. All of these species depend on one another and the coral itself for their survival. As in all ecosystems on Earth, organisms that live there maintain a delicate biological balance of competition and cooperation.

Reefs are busy centers of activity in an otherwise scantily populated ocean landscape. Their nooks and crannies provide hiding places, nurseries, and spawning grounds for many types of sea organisms. Each group of organisms that moves into a coral reef helps attract and maintain other kinds of living things. The mature reef hosts hundreds of species in a bright display of color and activity.

Coral reefs are small but invaluable pieces of the Earth's ecosystem. Reefs are more sensitive to pollution and other changes in environment than most other ecosystems are and are the first to reflect damage. Scientists watch them closely for signs of harm, knowing that what happens to coral reefs may eventually happen to other ecosystems.

2

Microbes and Plants
Simple Organisms and Algae on the Coral Reef

*F*ew places on Earth rival the abundance and splendor of life on the coral reef. A reef visitor can spot living things in almost every size, shape, and color; however, some of the most important reef inhabitants cannot be seen with the naked eye. These invisible organisms live on the reef floor or float in the water column, the huge expanse of water below the surface.

The organization of living things on coral reefs is unique. In most oceans, upper regions of the water teem with plankton, communities of tiny, drifting organisms. The plantlike members of this community, the phytoplankton, are able to carry out photosynthesis. The rest of the community is zooplankton, and it is made up of very small living things that cannot photosynthesize. In seas where the populations of plankton are substantial, waters are also rich in minerals and nutrients.

The waters around coral reefs are low in nutrients and have very small populations of plankton. It is this very lack of nutrients and plankton that make the waters of reefs so beautiful. Their vivid blue color is a reflection of the sky, and their crystal-clear transparency is due to the absence of living things in the water column.

Despite low levels of nutrients, coral reef waters are extremely productive parts of the oceans. Productivity refers to the amount of photosynthesis that takes place in an ecosystem, and therefore the amount of food created. Productivity on reefs is 50 to 100 times greater than in nearby ocean waters.

Several kinds of organisms contribute to the elevated productivity on reefs. Some of the primary producers are large algae, sea grasses, and sizable populations of microscopic algae. Many of these green, one-celled organisms live in the tissues of corals and a few other types of simple animals.

Food Chains and Photosynthesis

Living things must have energy to survive. In an ecosystem, the path that energy takes as it moves from one organism to another is called a food chain. The Sun is the major source of energy for most food chains. Organisms that can capture the Sun's energy are called producers, or autotrophs, because they are able to produce food molecules. Living things that cannot capture energy must eat food and are referred to as consumers, or heterotrophs. Heterotrophs that eat plants are herbivores, and those that eat animals are carnivores. Organisms that eat plants and animals are described as omnivores.

When living things die, another group of organisms in the food chain—the decomposers, or detritivores—uses the energy tied up in the lifeless bodies. Detritivores break down dead or decaying matter, returning the nutrients to the environment. Nutrients in ecosystems are constantly recycled through interlocking food chains called food webs. Energy, on the other hand, cannot be recycled. It is eventually lost to the system in the form of heat.

Autotrophs can capture the Sun's energy because they contain the green pigment chlorophyll. During photosynthesis, detailed in Figure 2.1, autotrophs use the Sun's energy to rearrange the carbon atoms from carbon dioxide gas to form glucose molecules. Glucose is the primary food or energy source for living things. The hydrogen and oxygen atoms needed to form glucose come from molecules of water. Producers give off the extra oxygen atoms that are generated during photosynthesis as oxygen gas.

Autotrophs usually make more glucose than they need, so they store some for later use. Heterotrophs consume this stored glucose to support their own life processes. In the long run, it is an ecosystem's productivity that determines the types and numbers of organisms that can live there.

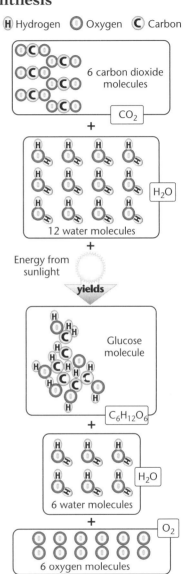

Fig. 2.1   During photosynthesis, the energy of sunlight is used to rearrange the components of carbon dioxide and water molecules to form glucose, water, and oxygen.

Simple Coral Reef Microbes

Cyanobacteria are the smallest and simplest producers on the coral reef. Producers, or autotrophs, are organisms that are capable of making food molecules. Cyanobacteria are members of the kingdom Monera and have been on Earth longer than any other living thing. Cells just like them are believed to have formed the reefs of ancient seas on early Earth. Cyanobacteria are still abundant on present-day coral reefs, although they play different roles than those of their ancestors. Some types of cyanobacteria provide life-sustaining food and oxygen for the coral reef system, but others are responsible for disease and death.

A few species of cyanobacteria are capable of nitrogen fixation, a job that falls to heterotrophs, organisms that cannot

Kingdoms of Living Things

There are millions of different kinds of living things on Earth. To study them, scientists called taxonomists classify organisms by their characteristics. The first taxonomist was Carolus Linnaeus (1707–78), a Swedish naturalist who separated all creatures into two extremely large groups, or kingdoms: Plantae (plants) and Animalia (animals). By the middle of the 19th century, these two kingdoms had been joined by the newly designated Protista, the microscopic organisms, and Fungi. When microscopes advanced to the point that taxonomists could differentiate the characteristics of microorganisms, Protista was divided to include the kingdom Monera. By 1969, a five-kingdom classification system made up of Monera (bacteria), Protista (protozoans), Fungi, Animalia, and

Plantae was established. The five-kingdom system is still in use today, although most scientists prefer to separate monerans into two groups, the kingdom Archaebacteria and the kingdom Eubacteria.

Monerans are the smallest creatures on Earth, and their cells are much simpler than the cells of other living things. Monerans that cannot make their own food are known as bacteria and include organisms such as *Escherichia coli* and *Bacillus anthracis*. Photosynthetic monerans are collectively called cyanobacteria, and include *Anabaena affinis* and *Leptolyngbya fragilis*. In the six-kingdom classification system, the most common monerans, those that live in water, soil, and on other living things, are placed in the kingdom Eubacteria. Archaebacteria are the inhabi-

produce food but must consume it, in many other ecosystems. In all cells nitrogen is an essential element that is used to make proteins and DNA, the genetic material that carries each cell's blueprint. There is plenty of nitrogen gas in the atmosphere and dissolved in ocean water, but the majority of cells cannot capture and use it. Cyanobacteria are one of the few organisms that can take in atmospheric nitrogen and change it into life-supporting nitrogen compounds.

In the reef cyanobacteria also supply nutrition for animals such as sponges that filter their food out of the water. In addition these monerans are captured and consumed by heterotrophic protists (protozoans) and small animals.

Under optimal conditions, cyanobacteria can reproduce rapidly, doubling their numbers within hours. Periods of fast

tants of extreme situations, such as hot underwater geothermal vents or extremely salty lakebeds.

Another kingdom of one-celled organisms, Protista, includes amoeba, euglena, and diatoms. Unlike monerans, protists are large, complex cells that are structurally like the cells of multicellular organisms. Members of the Protista kingdom are a diverse group varying in mobility, size, shape, and feeding strategies. A number are autotrophs, some heterotrophs, and others are mixotrophs, organisms that can make their own food and eat other organisms, depending on the conditions dictated by their environment.

The Fungi kingdom consists primarily of multicelled organisms, like molds and mildews, but there are a few one-celled members, such as the yeasts. Fungi cannot move around, and they are unable to make their own food because they do not contain chlorophyll. They are heterotrophs that feed by secreting digestive enzymes on organic material, then absorbing that material into their bodies.

The other two kingdoms, Plantae and Animalia, are also composed of multicelled organisms. Plants, including seaweeds, trees, and dandelions, do not move around but get their food by converting the Sun's energy into simple carbon compounds. Therefore, plants are autotrophs. Animals, on the other hand, cannot make their own food. These organisms are heterotrophs, and they include fish, whales, and humans, all of which must actively seek the food they eat.

growth can lead to population explosions, events that are also called algal blooms. In a bloom, the numbers of cyanobacteria increase so rapidly that the organisms form a dense blanket in the upper layers of the water. On a reef an algal bloom could block sunlight needed by photosynthetic organisms living deeper in the water. Low levels of sunlight slow reef productivity.

A few species of cyanobacteria also cause diseases in corals. Coral diseases are one of the primary reasons that living corals are lost to reefs. White band and red band are among several "band diseases" attributed to cyanobacteria. Black band disease, one of the worst and most extensive, is caused by a species of cyanobacteria named *Phormidium corallyticum*, although other microbes may also be involved. This particular disease can move across a reef quickly, killing it at a rate of a half-inch a day. In cases of severe infection, all of the coral animals may be destroyed.

Protists and Fungi

Protists are another group of one-celled organisms, although these cells are larger and more complex than the cells of monerans. There are many types of protists living on the reef. Some of the key players are diatoms, dinoflagellates, and foraminiferans.

Like most groups of protists, diatoms are a highly diverse group. They use all types of nutritional strategies, and their ranks include autotrophs, heterotrophs, and mixotrophs (organisms that are both autotrophic and heterotrophic). Some species are capable of rapid movement, while others are stationary during their entire lives. Despite their differences, all diatoms have some characteristics in common.

The fragile cells of diatoms are covered with protective shells, or frustules. Each frustule contains a large component of silica, the same material that is found in sand and glass; therefore, diatom frustules look like tiny glass hat boxes that are topped with lids. The silica shells of several marine diatoms are visible in Figure 2.2. These delicate silica structures are pierced with openings that permit the organisms inside to interact with

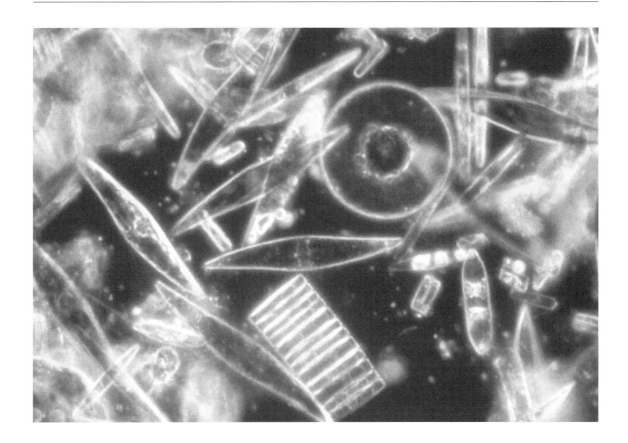

their watery environments. The geometric patterns created by openings in frustules are unique to each species and serve as a method of identification.

Diatoms create frustules in two basic shapes: round or elongated. Species that make round frustules generally inhabit phytoplankton, so they float in the water column. The elongated, or pinnate, organisms make their homes on the sand, in sea grass, and among algae. Most reef diatoms are elongated forms that inhabit the reef floor.

As in many protists, reproduction in diatoms is by binary fission, an asexual method. Asexual reproduction involves only one parent and forms clones, or exact duplicates, of the parent. Binary fission, however, poses a challenge to diatoms that cells of other species of protists do not face: The frustule must also divide, along with the cell. When a parent diatom splits apart, it forms two "daughter" cells. Each daughter

Fig. 2.2   Marine diatoms may be round or elongated in shape. (Courtesy of NOAA, Coral Kingdom Collection)

inherits one portion of the parent's frustule; one daughter gets the larger top portion of the frustule. The other receives the smaller bottom piece. The inherited portions become lids for each daughter cell, and both daughters grow new lower parts. The result is that one cell is the same size as the parent, and one is smaller. Over several generations, frustules become too small to divide further so the organisms leave their old, undersized shells. Afterward, the cells either undergo a period of growth, then secrete new, spacious frustules, or they go through sexual reproduction.

Sexual reproduction can occur in several ways. In some species, a small diatom cell breaks apart into little pieces, each of which swims around until it finds another diatom cell with which it can fuse. The product of their fusion builds the new frustule. In other species, two adult diatom cells line up beside each other. They divide, then exchange one daughter cell. The new pairs of daughter cells fuse, resulting in two new cells. In both cases, the resulting cells have genetic material from two parents.

Advantages of Sexual Reproduction

Even though asexual reproduction seems like a simple solution to continuing a species, many monerans and protists also undergo sexual reproduction. While asexual reproduction expands a population, it does not make it possible for the population to change in any way. All of the organisms created in asexual reproduction are clones, so they have the same genetic information and the same characteristics as the parent organism. As long as environmental conditions remain steady, asexual reproduction maintains a healthy population; however, if anything in the environment changes, the population may suddenly be at risk. Because all the individuals are alike, any problem that may befall one cell will probably visit them all, possibly resulting in the loss of the entire population.

In organisms that reproduce sexually, all of the offspring are different. Each one contains a unique set of genetic information, half of it inherited from one parent and half from the other. Since individuals in the population vary, it is unlikely that a change in the environment would create problems for everyone. In fact, a change that reduces the survival rate of some might improve the survival rate of others.

Most marine diatoms are autotrophs and are critically important producers in reef food chains. For this reason, green diatoms are nicknamed the "grass of the sea." The few species that cannot photosynthesize live in places that are rich in dissolved organic matter, which they absorb. Some species have the best of both worlds, photosynthesizing when light is available, absorbing nutrition when it is not.

On coral reefs, diatoms serve as meals for other organisms. Larger protists prey on them, and animals that graze sea grass and algae consume them as they feed on the plants. When conditions are good, growth of autotrophic diatom populations can lead to a diatom bloom. Although these blooms can be beneficial to the reef ecosystems by providing plenty of food for grazers, they often create problems. Some species of diatoms manufacture toxins that are designed to discourage predators. Many of these chemicals do not harm the shellfish that eat the diatoms, but they do accumulate in the bodies of those animals. When people eat contaminated shellfish, they can become very sick. One species of diatom causes a condition in humans called amnesic shellfish poisoning, with symptoms such as memory loss, disorientation, and coma. Severe cases can be fatal.

There are hundreds of species of dinoflagellates in the ocean, and several of them inhabit coral reefs. A dinoflagellate has two flagella, long, whiplike structures that propel the organism through the water. One flagellum fits into a groove that wraps around the cell, and the other sticks out behind the cell. Free-living dinoflagellates are covered with protective plates made of cellulose, the same material that forms the woody parts of plants.

Like most species of protists, dinoflagellates usually reproduce by binary fission. In the armored species, each daughter cell inherits half of the armor, then makes the other half. Occasionally, two cells will fuse to form a single large cell that contains the DNA of both parental cells. The large cell splits into two daughter cells, each with new combinations of DNA.

Dinoflagellates have developed several strategies for getting their nutrition. About half of the species contain chlorophyll as well as accessory pigments for photosynthesis. Accessory pigments are responsible for the colors of dinoflagellates, which range from golden brown to green. Most of the heterotrophic species of diatoms are colorless. They feed by engulfing tiny prey or by absorbing dissolved organic matter from seawater. Some species can get their nutrition in both ways, making their own food as well as absorbing it.

In the reef ecosystem, free-living dinoflagellates are prey to many organisms, such as larger protists, fish larvae, and small invertebrates. To discourage their predators, a few species of dinoflagellates use the same strategy adopted by other one-celled autotrophs: They produce potent toxins. *Gonyaulax tamarensis* is a dinoflagellate that makes a neurotoxin. As reef fish graze on seaweeds, they may consume. *G. tamarensis* living on the plants. With each feeding the toxin is stored in the grazer's fatty tissues, building up over time. Toxin levels are highest in the fattest animals, which are usually the most desirable catches. When these animals are eaten by large predators or by humans, the toxin can cause a condition called paralytic shellfish poisoning, whose symptoms include weakness, numbness, dizziness, and slow respiration. Death can occur from respiratory failure.

Another reef-dwelling species, *Gambierdiscus toxicus*, normally lives quietly on the surfaces of seaweeds in areas of the reef that are protected from waves. However, if predation is high, *G. toxicus* releases a chemical, ciguatoxin, that can poison fish, shellfish, or humans. Each year, tens of thousands of people worldwide are affected by the toxin, with a 1 percent mortality rate.

In the coral reef environment, some of the most important dinoflagellates are the species that reside inside the bodies of reef-building corals. By living together in a symbiotic relationship, both the host coral and their dinoflagellate guests benefit. The protists supply food and oxygen for the coral. The dinoflagellates provide most of the coral's nutrition. In addition, the presence of dinoflagellates helps corals form their protective skeletons. In return, the coral provides invaluable services for the dinoflagellates. They supply food and oxygen, protect them from predators, and provide a place to live that is located high in the water, near the sunlight. The relationship between dinoflagellates and corals is not unique; there are other organisms that also support microscopic producers. As a group, symbiotic one-celled autotrophs that live in the tissues of plants or animals are known as zooxanthellae.

In corals, dinoflagellates inhabit the tissues that line the digestive tract. Populations of these protists can be quite dense. With up to 50 tiny green organisms in each coral cell, there may be about 1 million cells per 0.16 square inch (1 sq cm) of coral flesh. Living as zooxanthellae, dinoflagellates lose their cellulose armor and flagella and take on a smaller, rounded shape.

Apparently, the relationship works out well for both organisms. The dinoflagellates are capable of moving out, but they rarely do. In experiments, scientists have removed dinoflagellates from their hosts and found that the little one-celled creatures regained their armor and flagella and flourished as free-living organisms. However, the corals did not fare as well. Without their houseguests, their rate of growth slowed considerably.

Foraminifera (forams) are protists that live in shells of different shapes and sizes. Some forams construct one-chambered shells, while others assemble shells that have multiple chambers of increasing size. In a number of species, shells are secreted by the protists, but in others they are constructed of sand and other particles that protists glue together. Many foram shells are large, reaching lengths of several inches.

Foraminifera live all around the coral reef. Some float in the water as single cells or in globules with other cells. Many species live in colonies that encrust rocks or other hard substrates, looking very much like patches of red film or paint. Another type of foram lives on the surfaces of seaweeds. When forams die, their shells settle to the floor. In some parts of the world, foraminifera shells make up 90 percent of reef sediment.

Forams play key roles in the food chains of reefs. They are not photosynthetic, so these protists dine on diatoms, bacteria, and even small fish eggs that they capture with extensions of their cytoplasm called pseudopods. Forams can also absorb dissolved nutrients in the water. Some species take in and nurture zooxanthellae, a strategy that assures plenty of food when prey is scarce. In turn, forams are prey to grazing and sediment-scouring animals such as worms, snails, and some fish.

Individual forams can live as long as two years, quite an advanced age for protists. When it is time to reproduce, cells shed their shells, then break up into many cell fragments. Each of these fragments grows a flagellum and swims away. There are two possible outcomes for these traveling cell parts: They can either develop directly into new forams or fuse with a similar cell fragment to form a cell.

Fungi, like bacterial decomposers, break down organic matter in the coral reef, releasing its nutrients for use by other organisms. As fungi grow on the bodies of dead or decaying organisms, they send out tiny filaments called hyphae. Each filament releases enzymes that dissolve the tissues of the dead organism. Fungi can then absorb these dissolved tissues.

A few species of fungi are capable of causing diseases in organisms that live on the reef. *Aspergillus sydowii,* a type of fungus related to the kind that grows on old food, has been the source of a tremendous amount of damage to sea fan coral. Although *A. sydowii* has existed for a long time, researchers are finding that it is becoming more lethal. There are two possible explanations for this change. The fungus may have mutated into a deadlier form, or the corals may be less resistant to disease.

Plants

Two important groups of plants in the coral reef are seaweeds, also known as macroscopic algae, and sea grasses. Both types of organisms are autotrophs. Along with the one-celled autotrophs, these marine plants support the food chains in the reef.

Compared to other marine ecosystems, the number and diversity of plants in coral reefs are relatively low. Small plant populations may be due to the fact that competition for space on reefs is very high, and corals often outcompete plants for the best reef locations. In addition, many of the reef animals

Light and Algal Coloration

Light is a form of energy that travels in waves. When the Sun's light arrives at Earth, it has a white quality to it. White light is made up of the colors red, orange, yellow, green, blue, indigo, and violet. The color of light is dependent on the length of the light wave. Light in the visible spectrum contains colors and has wavelengths between 0.4 and 0.8 microns (1 micron equals $\frac{1}{1,000,000}$ of a meter, or .000001 m; a micron is also known as a micrometer). Violet light has the shortest wavelength in the visible spectrum and red has the longest.

Light is affected differently by water than it is by air. Air transmits light, but water can transmit, absorb, and reflect light. Water's ability to transmit light makes it possible for photosynthesis to take place beneath the surface. All of the wavelengths of visible light are not transmitted equally, however; some penetrate to greater depths than others.

Light on the red side of the spectrum is quickly absorbed by water as heat, so red only penetrates to 49.2 feet (15 m). Blue light is not absorbed as much, so it penetrates the deepest, reaching 100 feet (33 m). Green light, in the middle of the spectrum, reaches intermediate depths. When light enters water that is filled with particles such as dirt and plant matter, as in an estuary, it takes on a greenish brown hue because it only penetrates far enough to strike, and be reflected from, the particles. In tropical water where particulate levels are very low, light travels much deeper before it reaches enough particles to be reflected back to the surface, so tropical water appears blue. Below 1,500 feet (457.2 m), no light is able to penetrate.

Because of the way light behaves in water, aquatic plants do not receive as much of the Sun's energy as do plants on land. To compensate, most species contain some accessory pigments, chemicals that are adept at capturing blue and green light. These accessory pigments provide the plants additional light and thereby help macroalgae increase their rate of photosynthesis. Some of these pigments mask the green of chlorophyll and give colors to macroalgae that are not usually associated with plants. Accessory pigments explain why seaweed occurs in shades of brown, gold, and red. Green algae contain accessory pigments, too, but they do not mask the color of chlorophyll as the pigments in other kinds of algae do.

are grazers, and they may hold down the size of seaweed populations; in experiments where grazers are removed from an area of reef, plant density increases dramatically. Another reason may be that coral reef waters are low in nutrients needed to support abundant plant life. Despite all of these hurdles, several species of green, red, and brown macroalgae, as well as grasses, flourish in the reef environment.

Green Algae

There are between 50 and 100 species of green algae that make their homes on coral reefs. A few species are covered with calcium carbonate skeletons very similar to the ones that protect coral animals. These plants—called coralline, or calcareous, algae—play special roles in building reef structure.

Coralline algae occur in a wide variety of shapes and sizes. Some look like masses of fine, threadlike filaments spreading over the reef and rock surfaces. The filaments are able to trap sediments and cement the particles together. In this way, coralline algae strengthen and support the coral reef structure. Even if a storm hits and many coral colonies are broken, coralline algae quickly bind the pieces back together.

Unlike the encrusting forms, other species of coralline algae grow upright. They, too, produce calcium carbonate for structural strength and protection. When these algae die, the limestone in their bodies is converted to sand. About 50 percent of the sand found on coral reef beaches originated in coralline algae.

One of the abundant coralline green algae belongs to the genus *Halimeda*, shown in Figure 2.3. Resembling an underwater cactus, *Halimeda* can be found in most areas of the reef community. Fronds of *Halimeda* are modified into flat, calcified segments, each 0.2 to 1.2 inches (0.5 to 3 cm) wide. These segments are joined by very short, uncalcified areas, and they give the plant a branching shape. For *Halimeda*, the calcium carbonate skeleton is an asset because it provides support for the plant while protecting it from predation.

Coralline algae such as *Halimeda* help build and repair the coral reef. When the algae's hard segments break off, they fall to the reef floor where they are tossed and rolled around by waves and currents. Eventually, the calcium carbonate skeletons erode into particles of sand, much of which washes to the floor of the lagoon and contributes to the sandy bottom. Some is carried by moving water to other parts of the reef and deposited there. The sand settles into cracks and crevices on the reef, filling them. Up to 25 percent of a coral reef may be contributed by *Halimeda.*

Like all green algae, *Halimeda* can reproduce sexually and asexually. Asexual reproduction is fairly straightforward: If pieces of *Halimeda* break off and fall in a quiet place, they will grow there, eventually developing into a new plant. However, *Halimeda's* style of sexual reproduction has an unusual twist. *Halimeda* produces male and female gametes, or sex cells, just like other species of green algae. These cells discharge into the water, where they fuse and form zygotes. On the night before *Halimeda's* gametes are ready to be released, the entire plant loses its green color. The only green structures left on the plant are masses of tiny spots scattered on the surfaces of the fronds. These spots,

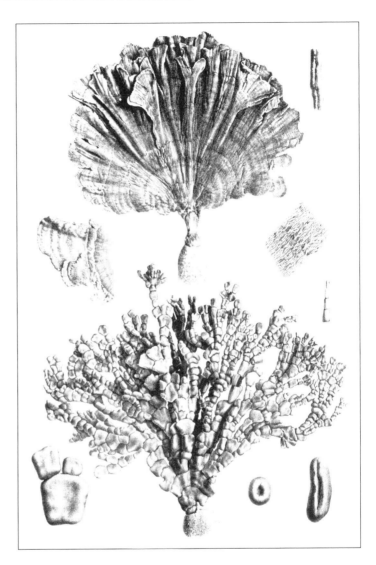

called gametangia, are the organs that make the sex cells. The contents of the plant's cells have moved into the gametangia, packing them with hundreds of chloroplasts, nuclei, mitochondria, and other cell parts. The next morning, the gametangia rupture, spewing flagellated gametes and cell contents into the seawater. The cells swim around until they encounter a gamete of the opposite sex. The two cells fuse to form a zygote that settles on the bottom and grows into a new *Halimeda* plant.

Fig. 2.3   Halimeda is a coralline green alga found on coral reefs. (Courtesy of NOAA, Coral Kingdom Collection)

Although *Halimeda* and many other species of green algae help build and support the coral reef, not all do. Some green algae actually damage the reef. The common green bubble alga, *Dictyosphaeria cavernosa*, is one of those. The fronds of *Dictyosphaeria* are modified into bright green bubbles that measure about 0.2 inch (5 mm) or larger in diameter. Patches of green bubble algae can form mats that measure 6.6 feet (2 m) across. These attach to the reef and grow across it very quickly. A mat of *Dictyosphaeria* can move over the reef at a rate of one inch (2.5 cm) a month. Green bubble alga is a fast grower because its mats form hollow chambers where gases and nutrients can get trapped. These raw materials then support the alga's fast rate of growth.

By overgrowing the coral, the algal mats smother the coral animals, destroying large areas of living reef. In addition, *Dictyophaeria* weakens the base of the reef structure. Eventually, it causes chunks of coral to break off, but it holds the coral loosely in place with its tendrils. When strong waves occur, they easily rip apart the mat and the coral supported by it.

Several species of another green alga, *Caulerpa*, are also found on reefs. These fast growing organisms occur in many varieties, some with delicate, fernlike fronds and others characterized by round, grapelike structures. *Caulerpa* can grow quickly by sending out runners that give rise to new plants. The runners cling securely to the reef floor with holdfasts.

Red Algae

Most red algae are multicellular plants that grow on rocks or other algae. Some of the most common red algae are encrusting forms that spread over hard surfaces, forming pink or red films on rocks and animal shells. Some other species of red algae develop tall plants with leafy fronds.

The encrusting coralline red alga *Porolithon* is especially noticeable on reef crests where it makes up much of the plant population. *Porolithon* is a tough plant that can endure occasional drying as well as intense wave action. Like the green alga *Halimeda*, *Porolithon* deposits calcium carbonate in its cell walls, giving it a rocklike texture. In all parts of the reef, this alga forms calcium carbonate that eventually finds its way into the reef. Calcium carbonate in *Porolithon* also acts as a mortar, cementing loose pieces of sand and rubble onto the reef structure. As a result, it improves the reef's overall strength and replaces much of the material that is lost to erosion.

Upright varieties of red coralline algae develop calcified branches that are jointed by uncalcified sections of frond. Tall red coralline algae are somewhat flexible and sway with the motion of tides and currents. Their hard cell walls protect them from damage by moving water and from grazing by predators.

Most species of red algae grow relatively slowly and have complex life histories that include phases of both sexual and asexual reproduction. Nori (*Porphyra*) are reef sea-weeds that look like limp, purple sheets. Nori uses a reproductive strategy called alternation of generations that is fairly typical of red algae. Both sexual and asexual reproductive structures are located on the thallus, or body, of nori. The asexual segments form spores that germinate to form new thalli. In the sexually reproducing sectors of the thallus, male and female cells release gametes that combine to form zygotes. These zygotes undergo cell division that results in spores. The spores settle onto the shells of animals such as crabs and barnacles where they grow into filamentous structures that bore into their hosts' shells. When the shell-boring phase of *Porphyra* matures, it then produces spores. Within these spores, cells undergo a special type of cell division called meiosis that reduces the amount of genetic material in each cell by half. These daughter cells settle to the bottom and grow into new thalli, bringing the process back to full circle.

In many types of algae, sperm are released into the sea, but female cells remain in the thallus. To ensure that the free-swimming sperm will find the eggs, some species use pheromones, chemicals that serve as signals. When eggs discharge their pheromones, sperm can follow them, like trail markers, to the eggs. Without pheromones, the sperm and eggs of some species would be unable to unite.

Differences in Terrestrial and Aquatic Plants

Even though plants that live in water look dramatically different from terrestrial plants, the two groups have a lot in common. Both types of plants capture the Sun's energy and use it to make food from raw materials. In each case, the raw materials required include carbon dioxide, water, and minerals. The differences in these two types of plants are adaptations to their specific environments.

Land plants are highly specialized for their lifestyles. They get their nutrients from two sources: soil and air. It is the job of roots to absorb water and minerals from the soil, as well as hold the plant in place. Essential materials are transported to cells in leaves by a system of tubes called vascular tissue. Leaves are in charge of taking in carbon dioxide gas from the atmosphere for photosynthesis. Once photosynthesis is complete, a second set of vascular tissue carries the food made by the leaves to the rest of the plant. Land plants are also equipped with woody stems and branches that hold them upright so that they can receive plenty of light.

Marine plants, called macroalgae or seaweeds, get their nutrients, water, and dissolved gases from seawater. Since water surrounds the entire marine plant, these dissolved nutrients simply diffuse into each cell. For this reason, marine plants do not have vascular tissue to accommodate

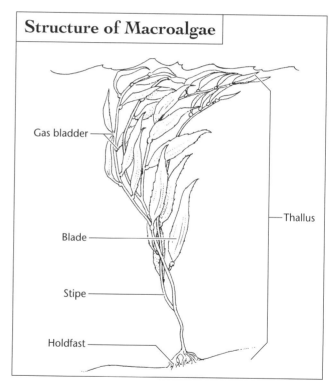

Structure of Macroalgae

Gas bladder

Thallus

Blade

Stipe

Holdfast

Fig. 2.4   The body, or thallus, of a macroalga is made up of leaflike blades, stemlike stipes, and rootlike holdfasts. Gas bladders on the stipes and blades help hold the plant near the top of the water column.

photosynthesis or to carry its products to each cell. In addition, marine plants do not need support structures because they are held up by the buoyant force of the water. Since water in the ocean is always moving, the bodies of marine plants are flexible, permitting them to go with that movement. Some marine plants secrete mucus to make their surfaces slick, further reducing their drag or resistance to water movement. Mucus also helps keep animals from eating them.

A plant that grows on land is described with terms such as *leaf, stem,* and *root.* Seaweeds are made up of different components, which are shown in Figure 2.4. The parts of seaweed that look like leaves are termed *blades,* or fronds. Some are equipped with small, gas-filled sacs, or *bladders,* that help keep them afloat and close to the sunlight. The gases in these bladders are usually nitrogen, argon, and oxygen. The stemlike structures of macroalgae are referred to as *stipes.* A root-shaped mass, the *holdfast,* anchors seaweeds but does not absorb nutrients like true roots do. Together, the blades, stipes, and holdfast make up the body, or *thallus,* of the macroalgae. Thalli take on many different forms, including tall and branched or thin and flat.

Brown Algae

A few species of brown algae live on the reef. Some are small and grow inconspicuously next to the corals, while others are prominent on the reef flat. Like many red algae, browns have complex life cycles that involve alternation of sexual and asexual generations.

Brown algae vary in form from encrusting growths to thin filaments, fleshy stalks, and giant kelps that reach sizes of 328 feet (100 m) long. Most brown algae cling to the bottom with holdfasts. Exceptions are the encrusting forms and a type referred to as sargassum, which floats on inshore reef waters.

Turbinaria is a reef brown alga that has a tough, spiny thallus capable of deterring would-be grazers. *Padina*, a fan-shaped brown alga, contributes to reef building. On the outer surface of its thallus, it forms bands of calcium carbonate that helps cement the reef together.

Sea Grasses and Mangroves

Although the green, brown, and red macroalgae play key roles in supporting the reef ecosystem, they are not the only large plants growing there. Two types of vascular plants, sea grasses and mangroves, are common around coral reefs. The ancestors of these true plants evolved to live on the land then moved back to the sea; therefore, they have many of the typical terrestrial plant adaptations such as roots and vascular systems. In addition, both mangroves and grasses form pollen and seeds. Once a year, grasses produce thousands of tiny underwater flowers that yield pollen grains that float in the water from one plant to another. After pollination, fertilized eggs mature into seeds that float away from the parent plants, sink, and start new beds of grass.

Besides reproducing sexually, sea grasses can spread asexually by sending out runners that generate upright shoots. In this way, they can quickly form large meadows on the sandy bottoms of lagoons. Sea grasses are not tolerant of intense wave action, so are usually found in the quieter regions of the reef.

Sea grass beds are important sources of food for many marine animals including shellfish, fish, and turtles. In addition, they provide good hiding places for the young of many species and often serve as nurseries where juveniles can find plenty of food and protection from predators.

Mangroves are large plants that are found growing on the beaches of island reefs. A mature mangrove tree can reach heights of 26.2 to 32.8 feet (8 to 10 m). Mangroves are never completely submerged; however, their root systems form extensive networks in shallow salt water. The presence of mangrove roots slows down the rate at which sediment-laden water swirls around the shore. When fast-moving water slows, it can no longer support suspended material, so the soil in it settles. The mangrove roots trap and hold soil, helping to stabilize and expand shorelines. The roots also provide shade, hiding places, and food for animals. Many species of reef-dwelling animals spend part of their lives among the roots.

Unlike sea grasses, mangroves produce above-water flowers. After pollination, egg cells develop into seeds on the trees. Seeds mature to seedlings before falling off the trees. If the tide is out when a seedling falls, it lands in the sediment and grows there. If the tide is in, the seedling drops into the water and is carried to a new location.

Conclusion

The organization of life on coral reefs was a mystery to scientists for many years. The first ecologists, scientists who study ecosystems, to visit coral reefs were surprised to identify only a few types of producers. Because phytoplankton in seawater supports most marine ecosystems, scientists expected to find it in large quantities around reefs. The lack of phytoplankton mystified them.

Today, scientists realize that the reef is rich in producers, but not in the form of phytoplankton. Chief among reef autotrophs are the zooxanthellae, green protists found in the cells of corals and a few other species of animals. Studies have demonstrated that zooxanthellae are responsible for at least half of the productivity in a reef system.

Coralline seaweeds are major contributors to the reef structure. In some reef ecosystems, they provide as much calcium carbonate as the coral animals. *Porphyra* is a typical coralline algae found on the reef crests. Although it is a slow grower, *Porphyra* can withstand a lot of wave action, drying conditions, and grazing by predators.

Several species of cyanobacteria and macroalgae are responsible for providing food for reef residents. The secondary pigments found in these organisms ensure their ability to trap the Sun's energy in the photosynthetic process, increasing their efficiency. All macroalgae are highly adapted for life in the sea, with flexible, slime-coated thalli that are secured to the ocean floor by holdfasts.

The reef producers, one-celled algae, macroalgae, and vascular plants, are an incredibly diverse group. All are highly adapted specialists capable of life in a harsh and highly competitive environment. This group of organisms is the key to the riches of coral reef ecosystems and the reason for their unrivaled success.

3

Sponges, Cnidarians, and Worms
Simple Reef Invertebrates

*T*here are millions of different kinds of animals in the world, yet they all have two basic characteristics in common. The first is that animals are multicellular organisms whose cells are organized according to their functions. In nearly all animals, groups of similar cells form tissues such as blood, muscle, and skin. Tissues are arranged into organs such as the heart, brain, and stomach. The second common characteristic is that all animals are heterotrophs. Animal tissues lack chlorophyll, so they are unable to use the Sun's energy to manufacture food. For this reason, animals must find and ingest food.

Of the millions of species of animals in the world, 95 percent are classified as invertebrates. As the numbers suggest, invertebrates form a highly successful group that has adapted to every niche of the environment. The principal characteristic of animals in this group is the absence of a backbone, a column of vertebrae around a central nerve chord. To support and protect their bodies, many invertebrates are equipped with hard, external skeletons.

On the reef, the statistics of invertebrate success hold true, and the greater part of reef animals are invertebrates. They include very simple creatures, such as sponges, corals, anemones, jellyfish, and worms, as well as more complex groups, such as clams, snails, octopuses, lobsters, and starfish. While the most primitive reef invertebrates are barely more than colonies of cells, the advanced ones possess organs as sophisticated as those of humans.

Sponges

Sponges are the simplest animals. A sponge has no nerves, bones, or other tissues and no organs, such as a stomach or heart. The body of a sponge functions a bit like a loose colony of cells, where each cell is capable of living independently. The basic anatomy of a sponge is elementary: Its body is made up of two layers sandwiched around a jellylike matrix called the mesoglea.

As shown in Figure 3.1, the outermost cell layer, the epidermis, acts as a protective skin, while the inner layer, or gastrodermis, collects and processes food. The gastrodermis contains specialized collar cells, also called chaonocytes, that are equipped with flagella. The mesoglea is interspersed with small skeletal needles termed spicules. Made of calcium carbonate, spicules look like slivers of glass when viewed under the microscope. Some species lack spicules, possessing instead fibers made of tough, rubbery protein called spongin. The job of both types of fibers is to brace up and support the sponge. Amoebalike cells move around in the matrix carrying out a variety of jobs, including arranging spicules, collecting and delivering food, and removing waste.

Sponges grow well in the unpolluted, shallow waters of the reef where they take on an endless array of shapes. Some have irregular forms, while others are shaped like flowers, vases, branches, or leaves. Most attach to the reef, a rock, or other hard surface. Sponges compete fiercely with other organisms for space. If a section of coral is damaged by a predator, a young sponge may settle in and take over the spot before the coral can heal itself. Unlike many of the algae, sponges also thrive in the very deep sections of the reef where fewer organisms are found. Since they do not require sunlight, water that is too deep for light to penetrate is not a problem for sponges.

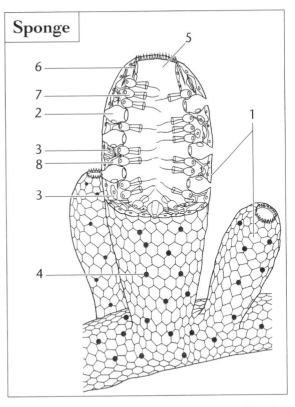

Sponge

Fig. 3.1   The epidermis (1) of a sponge is filled with tiny pores called porocytes (2). Amoebocytes (3) move around the sponge carrying food to cells. Water enters the sponge through an incurrent pore (4), flows into the central cavity, and exits through the osculum (5). Spicules (6) lend support to the sponge's body wall. Choanocytes (7) lash their flagella in the central cavity to keep water moving through the sponge and to gather bits of food that are suspended in the water. The mesoglea (8) is a jellylike matrix located between the epidermis and the cells that line the central cavity.

Sponges get their food by filter-feeding, a method of trapping tiny food particles that are suspended in the water. To ensure that plenty of food-laden water travels past them, chaonocytes inside the sponge lash their flagella back and forth. This action draws seawater through the thousands of tiny pores in the epidermis and into the gastrovascular cavity. Food particles suspended in the water adhere to the chaonocytes, which absorb the particles and digest them. Amoebalike cells then pick up the molecules of nutrients produced by digestion and carry them to all of the sponge cells. The water, minus its load of food, travels out of the sponge through the osculum, a large opening that works very much like the chimney in a fireplace.

To meet its nutritional needs, a sponge takes in a tremendous amount of water, handling up to five times its own volume every minute. Studies show that about 60 percent of the plankton around a reef is consumed by sponges. Much of what these animals eat is too small to be used by other filter feeders and would simply be lost to the ecosystem. As a sponge metabolizes all of this food, it produces and excretes waste products that serve as nutrient-rich fertilizers for many of the other reef organisms.

Sponges have two distinctly different phases of life. In the adult form, they are sessile, or attached to one place. When it is time to reproduce, some of the cells in each adult sponge transform into eggs and motile sperm. The sperm are released into water so that they can travel to another sponge where they are captured by collar cells and transported to the waiting eggs. The eggs and sperm unite to produce zygotes.

In some species, both eggs and sperm are released, and they unite in the water instead of inside a sponge. In either case, the zygotes develop into flagellated larvae. After swimming in the plankton for a while, larvae settle and begin growing. Sponges can also reproduce asexually. Small sponges grow from the base of an adult sponge by *budding*, a process in which a new organism grows from the body wall of the adult. Buds eventually break off and attach somewhere else. If a sponge is torn apart by waves, each part can develop into a new organism.

Because adult sponges are fixed in one place, they cannot run for safety when threatened by predators. For this reason, many sponges produce chemicals that discourage other animals from eating them. They also receive protection from their needle-like spicules that deliver painful sticks to nibblers. Many species of sponges avoid predators altogether by growing in crevices of coral where nothing can reach them.

A common and highly destructive type of reef sponge, *Cliona,* is capable of boring holes into the limestone skeletons of coral, causing the death of the coral animals inside. *Cliona* does not occupy the coral to feed from it; rather, its purpose is to live in the coral's space. Once established, these sponges can spread across a reef until all of the coral animals are dead. *Cliona* and other boring sponges are responsible for much of the bioerosion on reefs.

Other sponges play roles in helping build the coral reef community and providing shelter for several types of animals. For many small reef residents, a sponge is an ideal hangout. As predators avoid sponges that are armed with foul chemicals, any houseguest is also relatively safe. And because a sponge maintains a steady stream of food-laden water traveling through its body, the guest receives a constant supply of oxygen and nutrients.

The reef supports a variety of sponges, several of which resemble tubes or vases, as those shown in the upper color insert on page C-1. The giant tube sponge (*Aplysina lacunosa*) may be yellow, blue, or purple and usually grows in water 65 to 165 feet (20 and 50 m) deep. It forms long, hollow cylinders that are about 3 feet (1 m) tall and 3.9 inches (10 cm) in diameter. The strawberry vase sponge (*Mycale laxissima*), covered with fleshy, thick-walled cylinders that look like tiny urns, is usually less than 8 inches (20 cm) tall and found in waters below 33 feet (10 m). The beautiful iridescent tube sponge (*Spinosella plicifera*) gives off a glow in daylight.

Body Symmetry

An important characteristic of the body plan of an animal is its symmetry. Symmetry refers to the equivalence in size and shape of sections of an animal's body. Most animals exhibit body symmetry, but a few species of sponges are asymmetrical. If a plane were passed through the body of an asymmetrical sponge, slicing it in two, the parts would not be the same.

Some animals are radially symmetrical. Shaped like either short or long cylinders, these stationary or slow-moving organisms have distinct top and bottom surfaces but lack fronts and backs, heads or tails. A plane could pass through a radially symmetrical animal in several places to create two identical halves. Starfish, jellyfish, sea cucumbers, sea lilies, and sand dollars are a few examples of radially symmetrical animals.

The bodies of most animals are bilaterally symmetrical, a form in which a plane could pass through the animal only in one place to divide it into two equal parts. The two halves of a bilaterally symmetrical animal are mirror images of each other. Bilateral symmetry is associated with animals that move around. The leading part of a bilaterally symmetrical animal's body contains sense organs such as eyes and nose. Fish, whales, birds, snakes, and humans are all bilaterally symmetrical.

Scientists have special terms to describe the body of a bilaterally symmetrical animal, depicted in Figure 3.2. The head or front region is called the anterior portion and the opposite end, the hind region, is the posterior. The stomach or underside is the ventral side, and opposite that is the back, or dorsal, side. Structures located on the side of an animal are described as lateral.

Fig. 3.2   A sponge (a) is an asymmetrical animal. Starfish and jellyfish (b) are radially symmetrical; snails, turtles, and fish (c) are bilaterally symmetrical. In a bilaterally symmetrical animal, the head or front end is described as anterior and the tail end as posterior. The front or stomach side is ventral and the back or top side is dorsal. The sides of the animal are described as lateral.

Symmetries of Animals

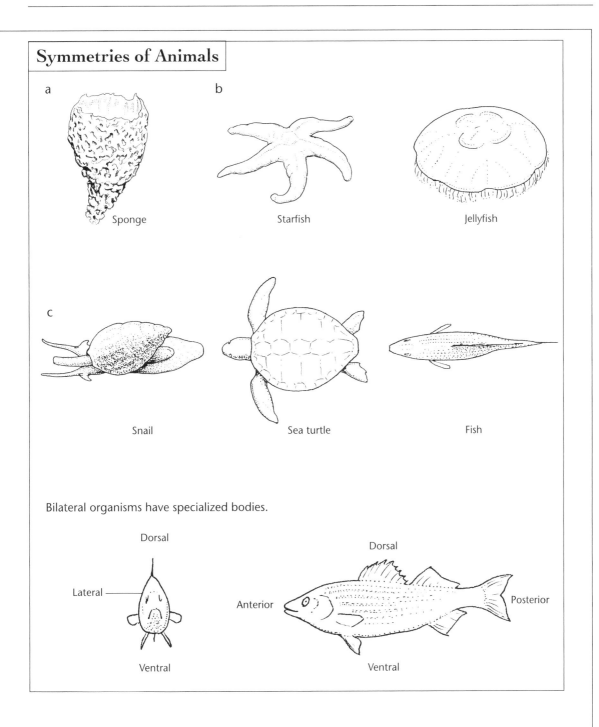

a

Sponge

b

Starfish

Jellyfish

c

Snail

Sea turtle

Fish

Bilateral organisms have specialized bodies.

Dorsal

Lateral

Ventral

Dorsal

Anterior

Posterior

Ventral

Cnidarians

Corals are the animals for which the coral reef is named and the primary architects of the system. Corals belong to a group of organisms called cnidarians, which contains some of the most beautiful and colorful animals on Earth. Hydrozoans, sea fans, anemones, and jellyfish are also cnidarians.

All cnidarians have a similar body plan, even though their bodies occur in two forms: a vase-shaped polyp or a bell-shaped medusa. As detailed in Figure 3.3, in both cases the body is a simple saclike structure with one opening, which is called a mouth. The sac is made of two cell layers, the epidermis and endodermis. Lying between them is the mesoglea.

Cnidarians are more complex than sponges. They have a network of nerves running through their bodies that enable them to respond to stimuli. If a cnidarian is touched, the nerve net relays a message to other cells, signaling them to contract and move away from the stimulus. The nerve net also helps a cnidarian locate, capture, and consume food.

Cnidarians are named for a type of specialized cell, the cnidocyte, that is unique to this group. Cnidocytes are stinging

Fig. 3.3 Cnidarians have two body plans: either a vase-shaped polyp (a) or a bell-shaped medusa (b). Each plan is equipped with tentacles (1), a gastrovascular cavity (2), and a single body opening, the mouth (3).

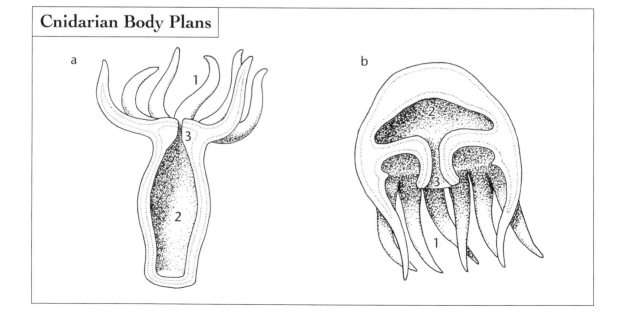

Cnidarian Body Plans

cells located in tentacles and used for protection and capturing prey. Each cnidocyte contains a nematocyst, a harpoonlike structure on a long filament. When something brushes against the cnidocyte, nerves send signals to the nematocyst triggering it to uncoil and extend the filament. In some species, filaments are tipped with barbs for injecting poison. In others, they are coated with a sticky material that holds prey. Once prey is snared, tentacles move it through the cnidarian's mouth and into the gastrovascular cavity where enzymes break it down, and cells in the gut cavity absorb the nutrients. Undigested parts are expelled through the mouth.

Corals can reproduce sexually and asexually. Between the ages of seven and 10 years, polyps become sexually mature and capable of forming male and female reproductive cells. Depending on the species, eggs may be fertilized internally or externally. Each zygote that results from fertilization develops into a small, swimming larva called a planula.

A planula is a ciliated larva about the size of the head of a pin that floats and swims some distance from its parents before settling. The larva attaches to a hard substrate and forms a polyp, never to move again. Once settled, a "founder" polyp starts making miniature copies of itself in the form of tiny buds that grow on all sides of the column. In this way, the original polyp begins a new colony of coral.

Corals blur the distinction between producers, consumers, and detritivores

Spawning and Brooding

For sexual reproduction to take place, male and female cells must come together. Many marine species spawn, or discharge one or both of their sex cells into the water. For this strategy to be successful, eggs and sperm must be released at the same time, which is why spawning usually occurs once a year at a specific time. Animals are cued to release gametes by specific environmental factors, such as the Moon's phase, length of daylight, or temperature.

The alternative strategy to spawning is brooding. Animals that brood release sperm into the water, while eggs remain within the mother. Sperm swim around until they find a female, enter her body, and fertilize the eggs. Eggs are brooded within the mother's body until time for them to hatch.

Fertilized eggs that are brooded have the advantage of protection from predators during development. In comparison, eggs fertilized in the water column or on the seafloor are at high risk from predators. For this reason, animals that brood their developing eggs only produce small numbers of gametes, while those that spawn discharge hundreds of thousands of gametes, a strategy that ensures that a few of the resulting offspring will survive until adulthood.

(organisms that feed on dead and decaying matter) because they use all three methods to meet their nutritional needs. This complex food plan helps corals survive by reducing their dependence on any one food source. Even if changes in the environment wipe out one supply of food, the corals have others on which they can fall back.

All corals are carnivores that can capture prey with their nematocyst-armed tentacles. Most species feed at night to avoid their own predators. In the dark, corals wave their tentacles in the water, waiting for small animals to swim within their reach.

Corals can also feed as detritivores. When levels of nutrients are high in the seawater, corals absorb dissolved organic matter (DOM) through their cell membranes. These molecules and compounds are used just like molecules from other food sources are, to meet energy needs and build cell structures. In addition, the tentacles of corals are coated with mucus so that tiny food particles that are floating in the water column will stick to them.

Many corals produce much of their own food with the help of zooxanthellae. Zooxanthellae are photosynthetic dinoflagellates that live within cells lining the gastrovascular cavity of coral. As zooxanthellae make their own food, some of it leaks into the coral host. Research on coral nutrition indicates that zooxanthellae meet the majority of coral nutritional needs.

Zooxanthellae perform another important job for the coral. The buildup of waste materials in an animal's tissues can slow down its metabolism and growth. Many simple marine animals must depend on the process of diffusion into the surrounding water to carry away these wastes. Zooxanthellae actively take up the waste products of their coral hosts. These products are the raw ingredients that zooxanthellae need to carry out photosynthesis.

Hard Corals

Some species of coral have hard protective skeletons around them, and others do not. Although both kinds are found on coral reefs, it is the hard corals that actually build the reef structure and are responsible for creating the reef environment.

Looking more like rocks or plants than animals, hard corals are polyps surrounded by protective skeletons. They vary in diameter from the size of a pinhead to more than 3 feet (0.9 m).

Coral skeletons are made of calcium carbonate, also known as limestone. Each individual polyp continually secretes a cup-shaped skeleton, or corallite, in which it sits. Throughout the polyp's life, the cup expands in height and diameter. With growth of the cup, the polyp is pushed up so that it always remains on the surface of the skeletal mass. Polyps often grow on top of large colonies, each new generation building atop the previous ones. The skeletons of the entire colony are called a corallum.

To construct their skeletons, corals absorb calcium ions from the seawater. Within their cells, they assemble these ions into crystals of calcium carbonate. Each newly formed crystal is held in the cell within a membrane-lined sac. Eventually, a crystal is pushed out through the cell membrane to the external surface. Once extruded to the coral's exterior, the crystal acts as a nucleus for continued crystal growth. The process of gathering calcium ions and using them to make skeletons is very efficient, and the amount of calcium carbonate that coral forms is staggering. Each day, the corals within 1 square mile (2.59 sq. km) of a reef are capable of creating almost 16 tons of calcium carbonate.

The presence of zooxanthellae in corals speeds the rate at which calcification occurs. In experiments, scientists removed the zooxanthellae by keeping corals in darkness for several months. Results show that corals without zooxanthellae grow very slowly. When their zooxanthellae are intact, corals can grow 14 times faster in the light than in dark. Furthermore, corals grow faster at noon, when the sun is bright, than at other times of the day.

There are hundreds of species of hard corals, and they show tremendous variation in size, shape, and color. Most hard corals are assigned common names based on something they resemble; for example, one of the branched species, shown in the lower color insert on page C-1, is called elkhorn coral (*Acropora palmata*) because it has thick and solid antlerlike branches. Colonies of elkhorn coral prefer water in the shallow, wave-pounded regions of the reef. Elkhorn corals are broadcast spawners that reproduce in August or September. New colonies are fast growing, elongating their branches at a rate of 2 to 4 inches (5 to 10 cm) per year, and most colonies reach their maximum size in 12 years. Because elkhorn coral lives in high-energy areas, storms often break apart the colonies. This process, known as fragmentation, is an important part of the coral life cycle. Broken branches reattach to the reef and grow into new coral colonies.

Another branching variety, staghorn coral (*A. cervicornis*), lives on the upper reef slope and produces cylindrical branched colonies in blue, green, or cream colors. Finger coral (*Porites compressa*) forms branched, fingerlike skeletal structures. The tips of each branch are rounded or flattened, and the color of the coral can vary from light brown to yellow.

Many coral species are shaped like rocks. Boulder coral (*Montastrea annularis*) is a brown to green species that can develop into mounds up to 10 feet (about 3 m) high. The yellow porous coral (*P. astreoides*) forms small, bright yellow boulders with bumpy surfaces. Common brain coral (*Diploria strigosa*), featured in the upper color insert on page C-2, is a medium-sized hemisphere outlined with regular hills and valleys that resemble brain matter. Other species of brain corals include large-grooved brain coral (*Colpophyllia natans*), which is pale brown with green-tinted valleys, and depressed brain coral (*D. labyrinthiformis*), which has hills with distinctive shallow depressions.

Soft Corals

Even though they do not contribute to reef building, soft corals are important members of the coral reef system. They are close relatives of hard corals, differing primarily in their lack of a hard external skeleton. The size of individual soft coral polyps rarely exceeds 0.2 inch (5 mm), but they live in colonies that grow up to 3 feet (about 1 m) tall. Despite the fact that they lack calcium carbonate skeletons, soft corals do not lack support tissue. In their mesogleal layer, they contain collagen fibers (similar to the material that forms tendons and ligaments in mammals) that enable these colonies to maintain a plantlike shape. Many of the soft corals live in areas of the reef that experience wave action and turbulence, and their flexible structure permits them to bend with the water movements. In other species, the internal skeleton includes some calcium carbonate spicules, which give these soft corals a slightly stiffer structure.

Soft coral animals are usually suspension feeders that gather their food with eight feathery tentacles. Tentacles capture bits of plankton and direct them through the mouth into the gastrovascular cavity. In soft corals, the gastrovascular cavity is divided into sections by eight walls of tissue. Each section is lined with the digestive cells. Most soft coral colonies have a system of internal tubes that distribute nutrients to all individuals.

To keep neighbors from infringing on their space, many species of soft corals protect themselves by producing chemicals that repel sponges and algae. Some soft corals also secrete toxins to ward off fish and other predators. Toxin-producing species display bright colors that warn other animals that they are dangerous and should be avoided. In general, the more fragile and easily harmed a soft coral species, the more predator-repelling toxin it contains.

Colonies of soft corals occur in a bewildering array of colors, including hues of red, blue, orange, and yellow. Many live on ledges where they extend into the flowing water to capture food. One common group, the leather corals (*Sarcophyton*), develop distinctive mushroom shapes. The spicules in these corals are densely arranged and mixed with collagen, making them tough enough to withstand the fastest waters without damage. Thick stalks anchor colonies of leather corals to the bottom, and polyps extend from broad caps at the anterior end of the stalks. When they are not feeding, the polyps pull back into the caps for protection. Leather corals are usually located in energetic areas like reef drop-offs and along the walls of channels.

There are hundreds of other species of soft corals. Deadman's fingers (*Briareum asbestinum*) grows in vertical stalks that may be 1.5 inches (3 cm) thick and 6 inches

Fig. 3.4   Sea fans are soft corals that have delicate, fanlike shapes. (Courtesy of Getty Images)

(15 cm) tall. When its polyps are extended and feeding, the corals resemble brown, fuzzy branches. Carnation coral (*Dendronephthya*), pictured in the lower color insert on page C-2, appears as soft and delicate as the flower for which it is named. Several species of soft corals are known as sea fans because of their characteristic flat, fan-shaped branchlets, as shown in Figure 3.4. Sea plume (*Muriceopsis flavida*) forms featherlike colonies very similar to those of the sea feathers(*Pseudopterogorgia acerosa*).

Hydrozoans

Hydroids, close cousins of the soft corals, are widespread reef inhabitants. Like hard corals, they build stony skeletons and live in colonies. Hydroids also feed very much like corals, depending on their zooxanthellae as their primary sources of nutrition. In addition, hydroids are suspension feeders as well as carnivores that capture their prey with stinging tentacles.

Several characteristics distinguish hydroids from hard corals. Unlike corals, hydroids experience both a polyp and a medusa stage of life. Plus, individuals in hydrozoan colonies specialize and divide the labor needed to support them, with some individuals taking charge of reproduction while others are involved in feeding, defense, and other functions.

Hydroids form three kinds of colonies: plantlike, coral-like, and jellyfish-like. The plant- and coral-like species can often be found on reefs. The plantlike colonies resemble slender stalks topped with multiple branches. One example, *Stylaster elegans,* grows to about 9 inches (22.75 cm) in length and is commonly found under coral overhangs.

The coral-like species produce colonies that are covered with skeletons of calcium carbonate. Their skeletons are so similar to hard corals in appearance that they have earned the nickname "false corals"; however, the two types are significantly different. Instead of growing in cup shapes, as true corals do, false coral form smooth skeletons that are dotted with tiny holes where the polyps are housed. Their ability to form skeletons makes coral-like hydroids important reef builders. In many systems they contribute a significant portion of limestone to the reef structure.

One group of false corals is well known to scuba drivers. The fire corals (*Millipora*) can deliver a painful sting to a fish or any other animal that touches them. Skeletons of *Millipora* may be purple, yellow, brown, or green, and they have white edges or tips on them. As in all false corals, the pores in the skeletons of *Millipora* house two types of polyps, feeding polyps and stinging polyps. Feeding polyps remain safely retracted within the skeleton until a meal is near; however, stinging polyps, which look like fine hairs waving in the water, are extended much of the time. These nematocyst-laden tentacles defend the colony, and are always poised and ready.

Anemones

Some of the most colorful and delicate reef invertebrates are the anemones, the so-called flowers of the sea. Like corals, anemones are cnidarians that only exist in the polyp stage. They have typical saclike, cnidarian bodies as well as nematocyst-laden tentacles around their mouths. Anemones do not build skeletons as their coral relatives do.

Anemones may occur in colonies or singly. The body of an anemone polyp is a thick column with two distinctive ends, both clearly visible in the upper color insert on page C-3. The upper end, or the oral disk, consists of a ring of tentacles around a narrow slit at the center, the mouth. Grooves beside the mouth bring in water continuously to provide oxygen to interior tissues. Depending on the species, oral disks vary in diameter from a fraction of an inch to 12 inches (30 cm).

The other end of the animal, the pedal disk, attaches to hard substrates. Some secrete a sticky adhesive to help hold them in place. Although the animals are classified as sessile, or sedentary, they can move around very slowly. They may shuffle across the reef floor on their pedal disks, similar to the way a snail moves on its foot, or somersault on their tentacles from one place to another. Some release their hold on the seafloor, inflate their bodies with air, and float to the surface, where they will rest for a while with their heads down.

Several species of anemones live on reefs. The maroon anemone (*Actinia bermudensis*) has short red tentacles bordered by bright blue warts. The rock anemone (*Anthopleura krebsi*) has greenish-yellow tentacles and rust-colored warts. The stinging anemone (*Lebrunia dana*) is duller in color, usually brown or white, but its tentacles deliver powerful stings that quickly kill prey and can injure humans.

Anemone Symbiotic Relationships

The tentacles of an anemone provide homes to a very select group of animals. The ringed anemone (*Bartholomea annulata*) and giant Caribbean anemone (*Condylactis gigantea*), for example, allow cleaning shrimp to live among their tentacles. These tiny transparent boarders remove bits of debris and parasites from anemones. The shrimp may even offer their cleaning services to fish, but if visiting fish show any aggression toward the hosting anemones, the shrimp will defend them fiercely.

The tricolor anemone (*Calliactis tricolor*), meanwhile, often agrees to a relationship with a hermit crab. The crab approaches the anemone and gently touches it, causing the anemone to retract its tentacles. The crab waits until the anemone relaxes and opens, then picks up the anemone and repositions it closer to its own shell. The anemone slowly climbs onto the dorsal side of the crab shell, where it will remain indefinitely. This arrangement protects the crab that now has a hat of stinging cells. The anemone benefits, also, since it has an opportunity to travel and find new sources of food.

Several kinds of fish, including species of clown fish such as the one shown in the lower color insert on page C-3, live with anemones in relationships that help both animals. In this partnership, a clown fish hides within the anemone's tentacles when it feels threatened and dines on bits of food that are left over after the anemone feeds. In return, clown fish protects the anemone from polyp-eating fish. The wastes produced by the clown fish supply the polyp with life-sustaining nitrogen and phosphorous compounds.

Jellyfish

Although relatively sparse, a few species of jellyfish can be found in the waters of the coral reef. A jellyfish is a cnidarian that spends its entire life in the medusa stage with its tentacles and mouth hanging down. To move in the water, jellyfish contract and expand their umbrella-shaped bodies, actions that thrust the animals from one place to another

The most commonly found jellyfish at a reef is the moon jelly (*Aurelia aurita*). Its bluish body, which measures about 10.5 inches (26 cm) across, is transparent, revealing cloverleaf-shaped reproductive organs inside. One to four long stinging tentacles hang down among several relatively short feeding ones. The long tentacles are armed with nematocysts that can deliver powerful stings.

Worms

Marine worms are one of the most abundant forms of life on the reef. Worms show a remarkable degree of adaptability and have evolved to occupy many reef niches. Despite their abundance, marine worms are not very easy to spot. The majority hide under rocks or in crevices, although a few species are free swimming. Worms get their food in a variety of ways, grazing on algae, preying on other animals, and scavenging dead and decaying material.

Around the reef, worms fall into one of two primary groups: flatworms and segmented worms. Of the two types, flatworms are the simplest and most primitive. Most marine flatworms are very thin, delicate animals. Flatworms have a simple digestive system with only one opening, so undigested matter is regurgitated out the mouth. To feed, a flatworm extends its pharynx, a muscular tube, onto its meal. The tube secretes digestive enzymes on the food, creating a soupy mix of partially digested tissue that can be sucked into the intestine.

Flatworms of the family Polycladida can be found in reef waters. These worms swim in undulating waves by throwing the sides of their bodies back and forth. Their name, which means "many branched," describes the highly divided digestive systems, which can be seen through the worms' transparent skins.

Polyclads have many would-be predators, especially among the fish and crustaceans. As a result, they have evolved several strategies to avoid predation. For example, most species feed at night when it is more difficult for predators to see them. In addition, many polyclads are camouflaged to blend into their environment. Others contain chemicals that either taste bad or are poisonous to their attacker. To advertise their distastefulness, flatworms with toxins, such as the specimen featured in the upper color insert on page C-4, sport bright colors and bold patterns. Using color to warn predators to stay away is a defensive strategy called aposematic coloration.

Polyclad flatworms are hermaphroditic, having both male and female reproductive organs. To cross-fertilize their gametes, two worms copulate, each donating sperm to the other. Sperm is exchanged through hypodermic-like stylets that each worm stabs into the body of its partner. The stylets are sharp, and stabbing sometimes causes damage to tissue, but injured worms usually recover within 24 hours. Once a stylet is injected, sperm are released to swim to the eggs. Fertilized eggs are laid on the reef floor under rocks or rubble and near a source of food. The eggs develop into larvae that swim for a few days before settling back to the reef floor where they become adult flatworms.

Most of the marine flatworms living around reefs have very poor eyesight. To get information about their environments, the worms depend on folds of tissue at the ends of their bodies. These folds create tentacle-like structures that can detect chemicals in the water, helping the worms navigate and find their food.

Although flatworms are common on the reef, they are greatly outnumbered by segmented worms. Reef segmented worms are colorful and flamboyant compared to their terrestrial counterparts. The most common types of segmented worms in marine environments are the polychaetes, or bristle worms. On some reefs, scientists have found that as many as 75 percent of the animal species are polychaetes. Their large bristles, structures called setae, are found on each body segment. Coral reef polychaetes are usually found living on the bottom in sediment, sand, or reef rubble.

Many reef polychaetes have a crown of true tentacles around their heads that are called radioles. These structures strain plankton from water and are the sites of gas exchange. Radioles are arranged as inverted funnels so that they can direct food toward the worm's mouth. Cilia constantly move the water around the radioles, creating currents that sweep food along.

One polychaete species, the sponge threadworm (*Haplosyllis spongicola*) bores into sponges, piercing them with tunnels. Sponge threadworms live in these tunnels where they are out of sight of predators and free to feed on the sponge tissue. Another group of polychaetes bore into corals, both dead and living. These worms eventually weaken the coral structure and cause sections of it to crumble.

Most species of segmented worms have separate sexes, although some are hermaphrodites. Those that spawn often do so when the corals spawn. Some segmented worms develop long structures called swimming lobes that help them swim up toward the surface during spawning. As the adults near the top of the water column, their bodies rupture, freeing eggs and sperm.

Two tube-dwelling polychaetes are the feather duster tube worm, as seen in the lower color insert on page C-4, and the Christmas tree worm. The Christmas tree worm (*Spirobranchus giganteus*), a Goliath in the world of polychaetes, lives in a tube that it builds on the surface of coral. As the worm grows, coral begins to surround the tube and eventually encloses it completely. To maintain contact with the outside world and avoid being sealed inside, the worm must continually lengthen its tube. When the tube is encased in the coral, only the worm's radioles are visible. The body, which is embedded in the coral, may be nearly 3 feet (1 m) in length. To feed, the worm sticks out its head and spreads it radioles, which resemble Christmas trees. Radioles can be yellow, pink, red, blue, white, gray, or brown, depending on the species. Their sticky mucus helps hold food particles, and their upside-down funnel shape directs food toward the mouth. When danger threatens, the worm can pull its head and radioles down into its tube.

Fireworms (*Hermodice carunculata*) are free-living polychaetes that grow to 12 inches (30 cm). Sensitive smelling

Palolo Worm

One species of polychaete worm—the *Eunice viridis,* or palolo worm—is cause for celebration in Samoa. Palolos are about 12 inches (30.5 cm) long and live in burrows in the coral reef. Their bodies are made up of two different kinds of segments. The anterior section contains the worm's organs, while the posterior section, called the epitoke, is a string of segments that contain gametes. One or two nights a year in October or November, worms crawl out of their burrows on the reef floor, and their epitokes break off. The pink and blue-green epitokes swirl to the surface like thousands of strands of colored spaghetti. Each segment is equipped with eyespots that are sensitive to light, ensuring that they move toward the sparkling light at the water's surface. During the night, the epitokes dissolve, and by morning the egg and sperm are released.

Natives of Samoa have a celebration and feast on "worm night." Using lights to attract the epitokes, they wade out into the surf and scoop up the worm segments. The worms are regarded as a delicacy, very much like caviar, and many choose to eat them raw on the spot. Some celebrants prefer to save their worm segments until the next day, when they can cook them in butter or bake them in a loaf of bread.

organs on their heads resemble pleated cushions. Fireworms are hunters that move boldly around the reef, protected by venom-filled bristles. Because the bristles can break off and embed in an attacker, they cause a long-lasting, burning sensation, hence the worm's name.

Fireworms are vicious predators of coral animals. They consume the polyps by spreading digestive enzymes over them, then sucking up the partially digested slush. A group of feeding worms leaves behind a trail of dead, white coral skeletons. Their favorite food is the tip of a staghorn coral, which one worm can devour in just a day. The feeding activity of fireworms can permanently change the reef because once the corals are killed, they rarely recover. In most cases, algae or sponges overgrow the area, preventing the return of coral animals.

Conclusion

Several groups of invertebrates construct their own quarters: Ants, for example, create hills, and bees build hives; however, no other group of animals builds a more dramatic home for itself than the corals. A coral reef, which resembles a pile of rocks, is actually an invertebrate superstructure made up of a living colony and the skeletons of their dead ancestors. The colony is always growing, but at a painfully slow rate of an inch or two (2.54 to 5 cm) each year. Reefs that are miles long were formed over thousands of years.

Chief inhabitants of the reef are the corals themselves. These simple invertebrates live alongside sponges, worms, anemones, hydroids, and thousands of other types of animals. Coral polyps wave their stinging tentacles in the water all through the night to grab any potential food item that comes their way. In addition, they stock their gastrovascular cavity with millions of zooxanthellae, tiny dinoflagellates that photosynthesize in the daytime.

A coral reef may support as many sponges, in both population size and diversity, as it does coral animals. Sponges, the simplest kinds of animals on Earth, are efficient filter feeders that continuously gather food. They grow in the company of colorful soft corals that resemble brightly colored plants, giant feathers, or ornate fans.

Close cousins to the corals are the hydroids, colonial cnidarians. Some of these sessile colonies form flexible, branching stalks, and others build limestone skeletons similar to those of the true corals. Nearby, growing singly or in clumps, are the anemones, cnidarians that lack any type of body covering. Anemones hold out their delicate but poisonous tentacles during the day to gather food. Some live in symbiotic relationships with shrimp, small fish, or crabs.

Scattered throughout the reef are the worms. Flatworms can be identified by their thin, transparent bodies and highly branched digestive systems. These simple animals feed by secreting digestive enzymes on their food, then sucking it up. The more advanced worms, the polychaetes, either crawl around the reef in search of food or catch food particles in their halo of radioles. Many are brightly colored predators that carry dangerous toxins.

The color and beauty of the coral reef make it easy to forget that each day is a life-and-death struggle for the simple animals that live there. Every invertebrate is a potential meal for another animal and survives simply because it has evolved some method of protection from predators.

4

Arthropods, Mollusks, and Echinoderms
Complex Invertebrates on the Coral Reef

*I*n the sea, as well as on land, invertebrates dominate the animal kingdom. Sponges and corals fill the reefscape with bright colors and endless body forms. Worms hide in every crevice, crack, and safe spot they can find. As abundant as these animals are, they only make up a small portion of the reef invertebrate population. Living alongside them are myriad invertebrates that are more complex and sophisticated. Many of these have well-developed body systems, some of which are very similar to those found in vertebrates.

Simple invertebrates, such as sponges, satisfy all of their oxygen needs by absorbing the gas directly through their cells. The larger and more complex invertebrates do not have this luxury. Complex invertebrates need greater quantities of oxygen than can be provided by simple absorption. In addition, their bodies are often covered with waterproof external structures, further reducing the amount of contact they have with oxygen-rich water. These structural characteristics demand a more complex respiratory system and have led to the development of gills.

Gills are respiratory organs made of thin tissues that contain thousands of tiny blood vessels. Because they are tightly folded, gills pack a large surface area into a small space. As water flows over the gills, oxygen that is dissolved in the water diffuses into the bloodstream. At the same time, carbon dioxide dissolved in the blood diffuses into the water and is carried out of the body.

Arthropods

Of all the invertebrates on the Earth, the largest group is composed of arthropods. Worldwide, there are about 80,000

Advantages and Disadvantages of an Exoskeleton

More than 80 percent of the animal species are equipped with a hard, outer covering called an exoskeleton. The functions of exoskeletons are similar to those of other types of skeletal systems. Like the internal skeletons (endoskeletons) of amphibians, reptiles, birds, and mammals, exoskeletons support the tissues and give shape to the bodies of invertebrates. Exoskeletons offer some other advantages. Serving as a suit of armor, they are excellent protection against predators. Also, because they completely cover an animal's tissues, exoskeletons prevent them from drying out. In addition, exoskeletons serve as points of attachment for muscles, providing animals with more leverage and mechanical advantage than an endoskeleton can offer. That is why a tiny shrimp can cut a fish in half with its claw or lift an object 50 times heavier than its own body.

Despite all their good points, exoskeletons have some drawbacks. They are heavy, so the only animals that have been successful with them over time are those that have remained small. In addition, an animal must molt, or shed, its exoskeleton to grow. During and immediately after a molt, an animal is unprotected and vulnerable to predators.

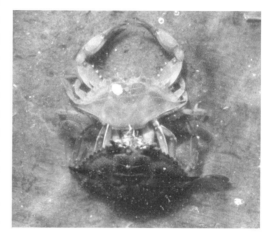

Fig. 4.1   A blue crab sheds its old shell so that it can grow; for a few days after, the crab is vulnerable to predators. (Courtesy Mary Hollinger, NODC biologist, NOAA)

species of these organisms. Many of the arthropods of the coral reef are familiar to nearly everyone and include animals such as shrimp, crabs, and lobsters.

Most arthropods are relatively small animals whose bodies are covered with hard, protective coats called exoskeletons. Exoskeletons give the bodies of animals structural support and protect them from predators. The tough skeletons of arthropods are primarily composed of chitin, an extremely hard, but highly flexible, material made of long chains of molecules that are similar in structure to cellulose.

Like the bodies of bristle worms and other segmented marine worms, arthropods' bodies are divided into sections. An arthropod has a definite head region that is specialized for handling food and gathering information about the environment. Most have compound eyes, which create multiple pictures and arrange them like tiles in a mosaic. Organisms with multiple eyes cannot focus on objects as well as human eyes, but they are very good at detecting motion.

Even though they are enclosed in a suit of formidable armor, arthropods can move about quickly. Their ease of motion is due to their jointed appendages. An appendage is a leg, antenna, or other part that extends from the main body of the animal.

Sexes are separate in arthropods, and mating is usually a seasonal event with elaborate courtship rituals. In many species the male deposits sperm in the female's body. The sperm are held here until eggs mature, then, as each egg leaves the ovary, sperm are released and fertilization occurs. Resulting zygotes mature into larvae that swim in the plankton for a short period of time before settling down on the reef floor to mature.

Crustaceans

The most common type of arthropod on the reef is the crustacean. Crustaceans include shrimp, lobsters, and crabs. As shown in Figure 4.2, the body segments of crustaceans are grouped into three specialized areas: head, thorax, and abdomen. Crustaceans have paired appendages attached to each segment that are used for walking, sensing, feeding, and defense. In a number of species, appendages form claws that are capable of exerting hundreds of pounds of pressure.

The biggest reef crustacean is the lobster. Even though the first pair of appendages are modified as claws in several lobster species, most reef lobsters are clawless. The appendages on the thorax of lobsters are adapted for walking along the reef floor and for swimming. Several lobster species make their homes on the reef. Two common ones are the spiny lobster (*Panulirus argus*) and the slipper lobster (*Parribacus antarcticus*).

The spiny lobster, as shown in the upper color insert on page C-5, is clawless, and its cylindrical body measures about

Crustacean Anatomy

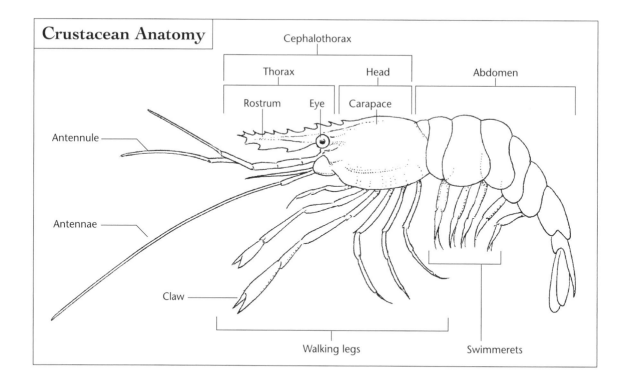

Cephalothorax

Thorax   Head   Abdomen

Rostrum   Eye   Carapace

Antennule

Antennae

Claw

Walking legs   Swimmerets

Fig. 4.2   The body of a crustacean is divided into three areas: head, thorax, and abdomen. The head and thorax are fused to form a cephalothorax.

1 foot (30 cm) long. For protection the lobster is covered with numerous, formidable sharp spines. Body color varies with age and sex, ranging from bright yellow to reddish brown and blue. The legs display yellow stripes, and the abdomen is covered in yellow spots.

The spiny lobster has a fascinating and complex, six-stage life history that has received a lot of attention from marine biologists. Studies show that this animal develops in several areas of the sea, including the open ocean, shallow coastal water, and coral reefs. During its 30-year life, a spiny grows from a miniscule larva to an adult of more than 11 pounds (5 kg), taking seven to 10 years to reach sexual maturity.

Generally, the mature spiny lobster hides in reef crevices, holes, and caves during the day, emerging at dusk to look for food in the sand and sea grass. A spiny will wander several yards from its den, preying on a variety of organisms including crabs, small fish, sea urchins, algae, and seaweed. During nighttime forages, lobsters may gather in small social aggregates. Before dawn, each lobster returns to the safety of its own, or a neighbor's, den. These nighttime wanderers can find their way around in the dark because they are very sensitive to the Earth's magnetic field and can use small variations in the field as landmarks. Their ability to find their dens is similar to the navigational senses of homing pigeons.

In the fall of each year, usually after a period of stormy weather, spiny lobsters congregate on the reef floor and begin a migration. They form a single file line called a queue, the head of one animal touching the tail of the animal in front of it, to march to deeper water. The lobsters travel day and night until they reach their destination. In the spring, they return to the reefs one by one. Scientists believe that these activities are related to water temperature and reproductive cycles.

Like the spiny, the slipper lobster is also clawless. The slipper lobster is drab or pale yellow and may grow to be 10 inches (25 cm) long. This carnivore also searches for food at night and will eat snails, oysters, clams, and the bodies of dead organisms. Slipper lobsters use their jaws to crack open prey. The bodies of these animals are flattened, a trait that helps them travel among the rocks unnoticed. Flat, shovel-like extensions on the head are the antennae.

Shrimp

Many crustaceans are classified as shrimp, animals that resemble lobsters but are much smaller. The term *shrimp* is not a scientific one but refers to a group of small crustaceans that have 10 jointed legs on the thorax and swimmerets on the abdomen. Unlike lobsters, shrimp are primarily swimmers instead of crawlers. Several species of shrimp live on the reef. The banded coral shrimp (*Stenopus hispidus*) and the painted dancing shrimp (*Hymenocera picta*) are two representative reef residents. Several species of mantis shrimp also make their homes there.

Mantis shrimps are longer than most other types of shrimp. These crustaceans, which are often vividly colored, are fierce and voracious predators. Depending on the species, their second pair of claws are modified into one of two forms: spears

Cleaning Symbiosis

The banded coral shrimp is one of several "cleaners" on the coral reef. Typically, a male and female pair of banded coral shrimp establish a cleaning station where they wait for fish. To let a fish know that they are cleaners, they signal their intentions with ritualized body language that includes waving their barber pole–colored bodies. A fish interested in the shrimps' services signals its willingness to be cleaned. These signals prevent cleaners from exposing themselves to fish that will eat them.

A busy cleaning station often has several fish lined up waiting for their turn. When a fish submits to cleaning, it may take unusual poses like headstands or exaggerated yawns so that the cleaners can do their work. The shrimp crawl all over the fish, removing parasites, mucus, and dead scales from their skin and mouth.

Cleaning is considered to be symbiotic behavior because the cleaner shrimps feed on the material that they remove from the fish. The fish in turn get "medical" treatment for parasites and skin problems. In studies where cleaners were removed from areas of the reef ecosystem, fish showed an increase in skin diseases. The relationship clearly benefits both organisms. Shrimps are not the only reef organisms that clean fish. Some small fish species, like gobies and wrasses, perform the same type of work.

or clubs. Those with clubs smash the legs of their prey to stop them, then crack open their bodies. This action is similar to the way a preying mantis attacks and accounts for the group's common name. The types that have spear-shaped claws use them to stab fish.

Mantis shrimps known as snapping, or pistol, shrimps have one enlarged claw that can snap shut with enough force to sound like a pistol shot. This unique snapping behavior serves two purposes: to warn trespassers to stay away or to stun prey, slowing them for a moment so that the shrimp can grab them.

Depending on the variety, pistol shrimps may have symbiotic relationships with fish, sponges, corals, or anemones. One species sets up housekeeping with gobies, small fish that are less than 4 inches (10 cm) long. In this association, the shrimp, which is an accomplished digger, creates and maintains a

Social Shrimp

One species of pistol shrimp forms colonies inside sponges, very much like bees in hives or ants in hills. These shrimps are considered to be eusocial, a term that describes groups of organisms in which most of the members are sterile workers while only one or two individuals are in charge of breeding. The interior of a sponge is similar to a chunk of Swiss cheese, riddled with tunnels and chambers. The pistol shrimps live in these passageways, gorging on sponge tissues that they scrape up with their small feeding claws. The predator-deterring toxins generated by sponges have no effect on these tiny boarders. They thrive in the sponge environment, often producing colonies of several hundred individuals. For the shrimps, the sponge is a perfect home, safe from predators and stocked with plenty of food.

In a colony of pistol shrimp, there is one breeding female. Often, but not always, one breeding male lives in the colony, too. All of the other members are males or juveniles who are not sexually differentiated. Many of these males patrol the sponge, defending it from other shrimp that would like to take over the space. This division of labor gives these shrimps a competitive edge and improves the chances of survival for all involved.

Eusocial shrimp colonies are possible because, unlike other crustaceans, the eggs of these animals do not hatch into larvae that swim in the plankton. Instead, they develop directly into shrimp that go straight to work in the colony. The young have no need to travel away from their parents in search of food and shelter. Except for the occasional juveniles who leave to strike out on their own, most of the snapping shrimps that are born in a sponge spend their entire lives there.

burrow in the sand. Because the shrimp has poor vision, it is in danger of wandering too far away from its burrow when predators are nearby. The sharp-sighted goby has no trouble keeping an eye out for danger, and it uses the shrimp's burrow as a base of operations. The goby spends its day hovering at the entrance of the burrow, occasionally dashing out to grab prey. When danger approaches, it backs into the den, warning the shrimp to do the same. The shrimp gets information from the goby by maintaining unbroken contact, constantly touching the little fish with its antennae and legs.

Crabs

Crabs are crustaceans, like shrimp and lobsters, but their bodies are more rounded. Crabs tuck their abdomens and tails up under their bellies. Most move by crawling along the ocean floor, although there are some species that swim. They have large front pincers that help them find and catch prey like clams, small fish, snails, and other crabs. Crabs do not stalk or pursue their meals; they usually wait for prey to swim by. They also scavenge dead and decaying organisms and look for food by sifting through sand or silt with their legs and antennae.

The spotted porcelain crab (*Porcellana saya*) is about 1 inch (2.5 cm) long. Its exoskeleton is orange red with white and violet spots that are ringed with red. Some species of spotted porcelain crabs are free living, but others live in symbiotic relationships with various types of hermit crabs and queen conchs (*Strombus gigas*).

Hermit crabs are not classified as true crabs. True crabs have five pairs of legs, including the claws, and hermit crabs have only four pairs. Like other crustaceans, a hermit crab has a crusty exoskeleton over most of its body; however, this protective armor does not extend over its abdomen, which is soft and vulnerable to predators. For this reason, hermit crabs have to find suitable shells to cover their abdomens. Without this protection, they could become a meal for some other animal.

To forage, hermit crabs travel across the reef floor, dragging their homes with them. Their shells are heavy, so the crabs cannot move very fast. To avoid some of their predators, they usually look for food at night. If frightened, hermit crabs retreat into their shells and cover the entrances with their large right claw. Their smaller left claw is for eating. The largest species in this group is the red hermit crab (*Petrochirus diogenes*), a rusty-red animal that can grow to 10 inches (25 cm). The red hermit crab is an active predator, but can scavenge food when necessary.

On the reef, true crabs include the lesser sponge crab (*Dromidia antillensis*), the purse crab (*Persephona punctata*), and the green reef crab (*Mithrax sculptus*). The lesser sponge crab is

Decorator and Sponge Crabs

Several types of crabs collectively known as the decorator and sponge crabs camouflage themselves from predators by carrying something on top of their shells. The toothed decorator crabs (*Dehaanius*) are small crustaceans with hooklike bristles on their backs where algae can be attached. Algae attached to a crab's shell acts as excellent camouflage. Like all crabs, these have five pairs of walking legs. The first pair is modified as claws, which pick up algae and place the material onto the hooks.

Sponge crabs, members of the Dromiidae family, hold their camouflage in place with two pairs of posterior legs that have been modified for grasping. The legs are turned up and armed with points that the crab sticks through the sponges to hold them in place. Sponges give off noxious chemicals that can cause predators to think twice before attacking.

When it is time to molt, a crab that is decorated in sponges lifts them off and sets them aside. When the new, larger shell is in place and hardened, the crab picks up the sponges and places them on its back again. Sponges do not grow as quickly as crabs, so after a few molts the crabs have outgrown the spongy hat. To solve this problem, crabs find new sponges, using their front claws to trim them for a proper fit. If a sponge is not available, the sponge crab will wear algae or anything else it can find.

only about 3 inches (7.5 cm) long with the last pair of legs bent over the back to hold a sponge. The purse crab is even smaller, at 2 inches (5 cm), with a white to gray shell that is marked with large red-brown spots. The extremely common tiny green reef crab is only 1 inch (2.5 cm) long and has hairy-looking legs.

Mollusks: Gastropods, Bivalves, and Cephalopods

The mollusks are a large group of arthropods that have a variety of outward appearances and include animals such as clams, octopuses, and snails. Because of their tremendous range of structures and styles, mollusks are divided into three groups: gastropods, bivalves, and cephalopods. Unlike their plainer terrestrial relatives, marine species of mollusks have extravagant forms that tend toward elaborately shaped shells and bright colors.

Even though the group is diverse, members share some common traits. The group's name, *mollusk*, literally means "soft bodied" and describes one of their primary characteristics. In addition to soft bodies, mollusks are also characterized by a foot that is used for locomotion. Internally, mollusk organs are covered with a thin tissue called the mantle. In some species the mantle secretes the shell and one or more defensive chemicals, such as ink, mucus, or acid.

A mollusk feeds with a file-like rod of muscle called the radula. This tongue-like organ is covered in sharp, rasping teeth that enable the animal to scrape up a variety of foods, including algae, animal tissue, or detritus. Most mollusks protect their soft internal parts with a hard shell. Like the body coverings of several types of marine invertebrates, the shells of mollusks are made of calcium carbonate. Mollusks exchange gases with the water through their gills.

Members of all three groups of mollusks live around the reef. Gastropods are common and include snails and nudibranchs. Snails glide over the sandy reef floor on one large muscular foot that is located in the center of their body. Snails' eyes, which are small, light-sensitive dots on their head, help them find food and keep watch for predators. Many snails have a spiral-shaped shell that contains and protects the internal organs. In some species, an operculum, a flap or door that can close the shell, protects the occupant from danger.

Most gastropods are hermaphrodites, individuals that have both male and female sex organs. Despite the convenience of having both sexes available, sperm are generally exchanged with another individual during mating. Eggs, which may be brooded in the mother's body, or laid as gelatinous masses or egg cases, hatch into shell-less larvae. Shells are produced by the mantle as the animals mature and are continually added to throughout life.

Several species of snails live on the reef. The stocky cerith (*Cerithium literatum*), which measures about 1 inch (2.5 cm) long, is white with rows of brown markings. In contrast, the tulip snail (*Fasciolaria tulipa*) has a smooth, reddish brown to gray shell that is covered with brown spiral lines. The soft tissues of this predatory animal are a dramatic red. One of the most beautiful gastropods on the reef is the flamingo tongue (*Cyphoma gibbosum*), shown in Figure 4.3. The flamingo tongue has a smooth pink and orange shell that is about 1 inch (2.5 cm) long and is enclosed in the snail's orange, leopard-spotted mantle tissue. One of the largest snails is the queen conch (*Strombus gigas*), a white-shelled herbivore measuring up to 12 inches (30 cm) long.

A group of colorful, shell-less gastropods also live on the reef. Although they have been nicknamed "sea slugs," these animals are not slugs but nudibranchs, so named because their gills (*branchia*) are nude (*nuda*). Most of the 3,000 species of nudibranchs are residents of the coral reef. In many species, gills are conspicuously displayed on their backs as knobby or feathery projections. Nudibranchs dine on an endless variety of foods, including sponges, tunicates (filter feeders related to

Fig. 4.3   The flamingo tongue snail has a brightly colored shell. (Courtesy of NOAA, Coral Kingdom Collection)

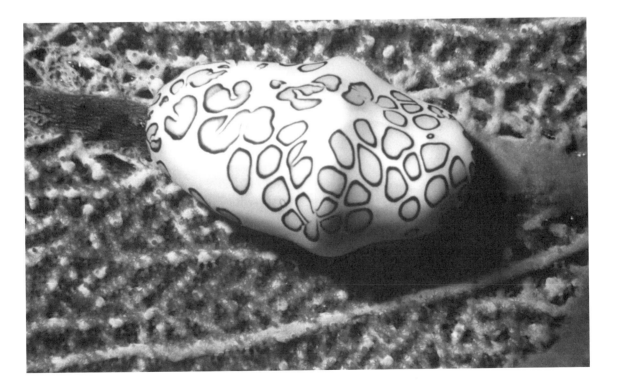

Weapons Recycled

Without shells to protect them, nudibranchs must rely on other strategies to discourage predators. The bodies of some nudibranchs produce poisonous or foul-tasting slime that nibbling fish find unappealing. Most of the poisonous species have bright colors that make the clear statement to predators: "I am poisonous—don't eat me!"

Some species of nudibranchs have the remarkable ability to disarm poisonous organisms and have even developed a technique for reusing their enemies' weapons for their own protection. Nudibranchs can dine on animals that other predators must avoid, including sponges, anemones, stinging coral, and jellyfish. When a nudibranch consumes an animal armed with nematocysts, the stinging cells pass through its digestive system unharmed and are transported to the dorsal surface and stored under the skin. Here the nematocysts continue to function, protecting the nudibranch from its predators.

vertebrates) anemones, corals, worms, crustaceans, and hydroids. A few species are able to consume algae without digesting them. Instead, their bodies store the algal cells just under the skin, keeping them alive and functional. The algae become solar-powered sources of food for the nudibranchs.

The spotted sea hare (*Aplysia dactylomela*) is a nudibranch that resembles a rabbit hunched and ready to hop. Its olive-drab body is 20 inches (50 cm) long and decorated with black ringlike spots. Structures that look like rabbit ears extend from its head, and a groove that is bordered with two flaps runs down its back. If this peaceful alga eater is disturbed, it emits a purple cloud to distract and confuse its predator.

Unlike the gastropods that have one shell, or are lacking a shell altogether, bivalves are animals that have two shells, or valves. The valves, which hinge together on one side and are opened and closed by strong muscles, provide these animals with protection from predators. Clams and oysters are some of the bivalves that live on the reef.

The foot of a bivalve helps it to either attach to a substrate or burrow into the sand. The foot is a large muscle that can be extended between the open shells. Although some gastropods scrape up food with a radula, bivalves use their gills to filter food from the water. A bivalve's gills, which are located in the mantle cavity, are covered with hairlike cilia and mucus. As water moves over the gills, tiny bits of food become trapped there. The cilia sweep the food into the bivalve's mouth.

Giant clams (*Tridacna maxima*) are extraordinary reef bivalves. Long ago mislabeled as "killer clams" because of their behemoth size, the largest individuals of this species measure 3 feet (1 m) wide and weigh a half ton. Giant clams, like all bivalves, are gentle filter feeders.

The much smaller file clam (*Lima*) is also found on the reef, measuring 1 to 3 inches (2.5 to 7.5 cm) long. Brown ridged shells protect the orange animals that live inside. Spiny, or thorny, oysters (*Spondylus americanus*) are reef residents that have round shells from which long spines stick in all directions. Below the spines, the shell surface is white, yellow, brown, purple, or red. The winged pearl oyster (*Pteria colymbus*) is a brownish purple animal that can be recognized by the long, thin extensions of its shell.

Cephalopods

The most intelligent and highly developed mollusks are the cephalopods. The name *cephalopod* literally means "head-foot" and refers to the fact that these animals have their feet, in the form of tentacles, in the head region. Members of this group that live around coral reefs include squid, octopuses, and cuttlefish.

Unlike bivalves and gastropods, most cephalopods have no shells, although squid and cuttlefish both have small, flexible pieces of shell material inside their bodies. In squid this material is called a pen, and in cuttlefish, a cuttlebone. A cuttlebone has chambers that can trap gases and give the cephalopod some buoyancy, while a pen is solid.

For protection cephalopods depend on their well-developed nervous systems and sense organs. Their eyes are extremely advanced and similar to human eyes. The primary difference is that the cephalopod focuses its eye by moving the lens back and forth, like a camera lens. In humans, the lens is focused by changing its shape.

A cephalopod body is designed for a free-living, predatory lifestyle. Octopuses are found singly, crawling across the bottom of the reef on their eight tentacles (see Figure 4.4). They spend a lot of time hiding in reef crevices and among rocks. Squid swim around in small schools, traveling head first with their 10 tentacles trailing behind them. Cuttlefish, which also have 10 tentacles, are most often found near the bottom of the

Cephalopod Camouflage

Cephalopods are soft bodied and have no external shells to protect them, so they are easy prey for hungry hunters like fish, sharks, and seals. Defense strategies used by cephalopods include techniques in camouflage that enable them to alter both the color and texture of their skin. As a cephalopod moves across the seafloor looking for food, the color of its skin changes almost instantly. This remarkable ability is mediated by the animal's advanced nervous system.

If startled, a cephalopod's eyes relay messages to the body telling it to go into a defensive mode. The eyes take in the color of the surroundings, then send nerve impulses to special skin cells called chromatophores that contain bags of pigment. When the bags expand, the color becomes intense; when they contract, the color fades to tiny dots. Camouflage is achieved by expansion of some chromatophores and contraction of others.

If there was a contest to judge the most creative use of chromatophores, the mimic octopus would win. The repetoire of this master of camouflage includes sea snakes, lionfish, and other poisonous animals. To imitate a lionfish, the cephalopod turns blue and flares its legs to look like poisonous fins. To impersonate a sea snake, the octopus changes its colors to yellow and black bands, tucks its body and all but two legs into a hole, then waves the two exposed legs in snake-like fashion.

reef. In all types of cephalopods, the tentacles encircle the head. These extensions are equipped with powerful suction cups for grasping prey. In the middle of the circle of tentacles is a mouth with strong, beaklike jaws.

In cephalopods, sexes are separate. The male has a specialized tentacle that transfers packets of sperm from his mantle cavity to the female's mantle cavity. After her eggs are fertilized, she lays chains of egg capsules on the reef floor or in caves. The female octopus broods over her eggs, blowing oxygenated water across them. During this time she eats little and dies after about 10 days. The eggs develop directly into juveniles, skipping the larval stage.

The blue-ringed octopus (*Hapalochaena lunulata*) is a lethal reef hunter. Its venom is located in saliva, so after biting its victim, the octopus spits venom into the wound. From above, the blue-ringed octopus is inconspicuously colored. However, when threatened, it flashes its bright blue warning rings.

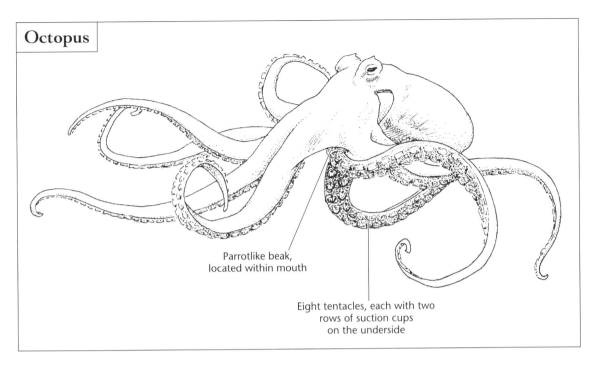

Octopus

Parrotlike beak,
located within mouth

Eight tentacles, each with two
rows of suction cups
on the underside

The Atlantic oval, or reef, squid (*Sepioteuthis sepioidea*) is 10 inches (25 cm) long with oval fins that begin near the mouth. Its yellowish skin is covered with purple dots. This squid swims with its short tentacles bunched, making them look like they are part of the head.

One species of cuttlefish found around the reef in the summer is the flashback cuttle (*Sepia latimanus*). This cephalopod comes into shallow, warm water to mate among the corals. In the daytime, the cephalopod stays near the bottom but becomes active at night. Its skirtlike fin encircles the body and helps stabilize the animal while swimming and hovering.

Fig. 4.4   An octopus has eight tentacles, each with two rows of suction cups. In the center of its body is a hard, parrotlike beak.

Echinoderms: Starfish, Brittle Stars, and Feather Stars

The familiar starfish (sea star) and its relatives are members of a group of spiny-skinned organisms, the echinoderms. Close relatives include brittle stars, basket stars, sea urchins, sea cucumbers, feather stars, and sea lilies. All of the animals in the starfish family have five or more arms radiating from a central body, shown in the lower color insert on page C-5. The mouth of an echinoderm is located on the underside and the anus on top of the body. Echinoderms may be carnivores, detritivores, or herbivores, depending on the species.

Tentacle-like structures called tube feet, visible in Figure 4.5, help echinoderms move slowly across the reef. The tube feet act like suction pads, grasping and releasing surfaces. A hydraulically controlled vascular system supplies water to each tube foot through small muscular tubes. As the feet press against an object, water is withdrawn, creating suction. When water is returned to the cups, the suction is broken, and they release.

To reproduce, echinoderms release sperm and eggs into water. These fuse to form zygotes that develop into larvae. After some time in the plankton, a larva settles to the bottom and

Fig. 4.5   On its ventral sides, a starfish, or sea star, has a centrally located mouth surrounded by five arms with tube feet. (Courtesy of Dr. James P. McVey, NOAA Sea Grant Program)

takes on typical echinoderm features. Most echinoderms can also reproduce asexually. If part of the animal breaks off, it may grow into a complete, new organism. All are capable of regenerating missing limbs, spines, and in some cases, intestines.

Starfish are common on the reefs. Most are carnivorous, feeding on sponges and small invertebrates. One species, the crown of thorns (*Acanthaster planci*), eats live coral polyps. Crown of thorns can reach 20 inches (50 cm) in width and may have 10 to 20 spiny arms. As is the case for all starfish, it eats by extruding its stomach over its prey, then drenching it in digestive juices. After the prey tissues liquefy, the crown of thorns pulls its stomach and the partially digested food back into the body.

A brittle star has a small central disk from which radiate five snakelike arms. Unlike starfish, brittle stars lack an anus, so waste products are eliminated through the mouth. On the underside of a brittle star, there are 10 slitlike openings at the base of the arm that function in breathing, as well as releasing eggs and sperm. Brittle stars often hide in the crevices in coral reefs, emerging at night to feed on plankton. If a predator happens to find and grab a brittle star, it has trouble hanging on. As its name suggests, this animal's arms break off easily, allowing for easy escape. A broken arm quickly regenerates.

Feather stars, or crinoids, are cup-shaped animals that have five to 200 feathery arms projecting upward from a central disk. The arms are coated with sticky material that helps them catch planktonic food. Feather stars cling to coral or rocks but can move by rolling, walking, crawling, swimming. They often establish symbiotic relationships with other animals including some species of shrimp, lobsters, and clingfish.

Sea Urchins and Sea Cucumbers

Sea urchins are echinoderms that look like rounded mounds of spines. Centrally located on their ventral surface is the mouth, which is equipped with jaws and teeth that scrape up algae. All sea urchins have tube feet and are able to move slowly across the floor of the reef.

The spines on different species of sea urchins show adaptations for unique environments. Urchins with thick spines use them to wedge their bodies between rocks, preventing predators from pulling them from their hiding places. Those that have flattened spines are able to tolerate the strong energy of waves. Sharp, venomous spines are adaptations to protect against predators. Some species have movable spines that help them crawl.

Savigny's sea urchin (*Diadema savignyi*) is a large, black reef species armed with sharp spines that can cause painful punctures. Mathae's sea urchin (*Echinometra mathaei*) is brown or pink with purple spines. The flower urchin (*Toxopneustes pileolus*) looks innocent, but its "flowers" are actually very small spines that are highly modified. Each little spine is a venomous pincer that carries enough toxin to cause paralysis or death in humans. The red pencil urchin (*Heterocentrotus mammilatus*) has flattened spines that help it wedge among rocks and survive in high-energy regions of the reef.

Sea cucumbers differ significantly from most other echinoderms. Instead of having the round or star shapes of their relatives, sea cucumbers, like the one in the upper color insert on page C-6, are cylindrical and look very much like the vegetable for which they are named. Some species are small and wormlike, but others reach impressive lengths of up to 3 feet (2 m).

During the day, sea cucumbers hide among rocks, but at night they emerge and work their way across the reef floor, "vacuuming" up sediment as they go. Inside their digestive systems, sand is filtered of its organic material, then the cleaned sand is excreted. The mouth, which is located at one end of the body, is surrounded by tentacles. Instead of sucking up sand, some species strain plankton from the water with their tentacles. The mouth does not contain teeth, but one species has teeth in its anus. In these animals, the anus also serves as the opening through which water is drawn into the body for gas exchange.

Sea cucumbers move sluggishly, crawling slowly on the sandy reef floor on their tube feet. These feet also enable sea cucumbers to cling very tightly to solid surfaces. Many divers who have tried to collect specimens have been disappointed to find that they will not budge. A few species are lacking tube feet and capable of swimming.

Of all reef inhabitants, sea cucumbers use one of the most unusual defense strategies. To protect themselves sea cucumbers project sticky threads of their intestines onto the attacker. The threads are coated with toxins that stop most predators in their tracks. When the threat has moved away, the intestinal structures begin to regenerate themselves.

The beaded sea cucumber (*Euapta lappa*) is a large one, measuring up to 3 feet (1 m) in length. Each thick, beadlike body segment is brown with thin yellow and black stripes. Beaded sea cucumbers have long, thick tentacles that are shaped like feathers. Members of this species do not have tube feet.

The five-toothed sea cucumber (*Actinopyga agassizii*), 1 foot (30 cm) long, ranging from gray to brown in color, is an unusual creature. The teeth of this echinoderm are located in its anus and can be seen when the sea cucumber "exhales" water that has been circulated over the gills. The anus is also the home of a commensal organism, the pearlfish (*Carpus*). In commensal relationships, one organism is benefited, and the other is not harmed or helped. In this case the adult pearlfish has a safe place to live inside the sea cucumber, which neither benefits nor is hurt by the pearlfish's presence. As juveniles, pearlfish are parasites, feeding on the sexual organs of their hosts, but as adults, they forage for food on the reef. For a foraging adult pearlfish to return to its host, it must swim into the cucumber's anus. The cucumber keeps its anal opening squeezed closed, so the pearlfish waits until the cucumber "exhales" and relaxes the opening, then the fish darts inside.

Conclusion

Coral reefs are the largest structures on Earth that are built and inhabited by invertebrates. Created by a few species of calcium carbonate–secreting organisms, a reef provides homes for thousands of other kinds of animals. The interactions between individuals in this highly diverse group of organisms regulate the ebb and flow of life in reef communities.

Marine invertebrates that live on land are relatively small animals, but some species in the sea grow to spectacular sizes.

The giant clam weighs up to a half ton, exceeding the size and weight of any terrestrial cousin. This startling difference in size is due to the advantages that marine environments have over terrestrial ones, especially for invertebrates with shells. The seafloor is a very stable environment, with few changes in physical factors such as light, temperature, and acidity. This constancy facilitates stable conditions within the animals' tissues. In addition, the water gives structural support to animals, reducing the consequences of weight and permitting animals to reach lengths that would be impossible to attain on land.

Reef invertebrates include animals that are classified as crustaceans, mollusks, and echinoderms. The crustaceans are typified by lobsters and shrimps, animals whose bodies have specialized segments. Shrimps show a tremendous diversity in lifestyle and have adapted for life in every niche of the reef. Their primary defense weapons are their claws, which are often modified for specific tasks.

Mollusks include gastropods, bivalves, and cephalopods. Gastropods such as snails and nudibranchs glide on one large, muscular foot. Snails live in twisted shells that house and protect their soft internal organs. Nudibranchs are without shells, and they depend entirely on their toxins and bright warning colors to keep predators at bay.

Bivalves, mollusks with two shells, such as clams and oysters, live quiet lives, sitting on the reef or hiding in the sediment. They are filter feeders that protect themselves from predators by snapping shut their thick shells.

The shell-less members of the mollusk group are the cephalopods, and on the reef they include the cuttlefish, squid, and octopus. Cephalopods are the intellectuals of the reef, capable of lightening-fast integration of sensory input. Octopuses are also masters of disguise that use their chromatophores to imitate deadly predators or to blend into the environment.

Although mollusks may be a little hard to spot on the reef, echinoderms are usually easy to find. Echinoderms have stiff, spiny skin that protects them from hungry predators. They travel across the reef floor, unhurried since most other animals cannot eat them. Echinoderms are capable of "walking"

▲ *Brown tube sponges are found on the coral reef of the Florida Keys.*
(*Courtesy of Florida Keys National Marine Sanctuary*)

▲ *Elkhorn corals can form large colonies on reefs.* (*Courtesy of Florida Keys National Marine Sanctuary*)

▲ *These brain coral (right) and sea fans are named for the objects they resemble.* (Courtesy of Florida Keys National Marine Sanctuary)

▲ *The tentacles of several carnation corals can be seen in this soft coral colony.* (Courtesy of Mr. Mohammed Al Momany, Aqaba, Jordan)

▲ *This stalked anemone has its tentacles fully extended.* (*Courtesy of Dr. James P. McVey, NOAA Sea Grant Program*)

▲ *The symbiotic relationship between this two-banded clown fish and sea anemone benefits both organisms.* (*Courtesy of Mr. Mohammed Al Momany, Aqaba, Jordan*)

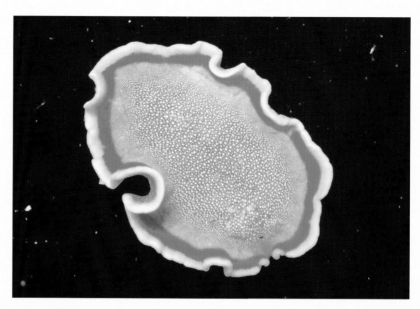

▲ *Many of the flatworms on coral reefs are brightly colored.* (*Courtesy of Dr. James P. McVey, NOAA Sea Grant Program*)

▲ *The feather duster tube worm extends its tentacles to capture food particles in the water.* (*Courtesy of Dr. Anthony R. Picciolo, NOAA NODC*)

▲ *The spiny lobster is a clawless species that hides in reef crevices and caves.* (*Courtesy of Dr. James P. McVey, NOAA Sea Grant Program*)

▲ *A marble sea star uses its tube feet to crawl across coral.* (*Courtesy Getty Images*)

▲ *Sea cucumbers are close relatives of starfish.* (*Courtesy of Dr. James P. McVey, NOAA Sea Grant Program*)

▲ *Red squirrelfish usually feed around the reef at night.* (*Courtesy of Dr. Anthony R. Picciolo, NOAA NODC*)

▲ *A moray eel hides in a crevice among the corals.* (*Courtesy of Mr. Mohammed Al Momany, Aqaba, Jordan*)

▲ *The spot on this spotfin butterfly fish confuses its predators, drawing their eyes away from the vulnerable head and eyes.* (*Courtesy of Jackie Reed, NOAA*)

▲ *Adult green sea turtles graze on algae growing among corals.*
(*Courtesy of Florida Keys National Marine Sanctuary*)

▲ *A humpback whale breaches in waters of the coral reef.* (*Courtesy of Sanctuary Collection, NOAA*)

slowly on multiple rows of tiny tube feet. Most prey on smaller invertebrates.

Crustaceans, mollusks, and echinoderms populate every part of the reef. Many are completely dependent on the reef's hiding places for their survival. A few are so well protected by their bodies' defense systems that they can exploit more open areas, like the reef floor, which is often covered in hard-shelled bivalves and spiny-skinned starfish. Because of their wide range of adaptations, reef invertebrate populations are large and rich.

5

Fish
A Rainbow of Colors

$\mathcal{T}$ he most active and obvious living things in the ocean are the fish, and the greatest diversity of fish species can be found on the coral reef. No other part of the marine environment can match the abundance and diversity of the reef fish populations. Some families of fish are adapted for life among the corals, while others are better suited to live among sea fans, sponges, and sea anemones. Many are specialized to live on the reef floor, while others makes their homes near the water's surface. The variations in reef fish are almost limitless.

Unlike the reef-building corals, armored crustaceans, and two-shelled clams, fish are vertebrates, animals with backbones. All vertebrates have internal skeletons for structure and support. The skeleton may be made of cartilage, a tough, flexible material; bone; or a combination of the two. Fish are vertebrates that have scales, fins, and gills.

The reef ecosystem can support large populations of fish because most of the inhabitants are specialized feeders. If fish competed with one another for food, many would be driven away by hunger. By specializing, reef fish have been able to develop clearly defined feeding strategies that allow each species to fill a specific role in the community that is slightly different from that of its neighbors'.

One strategy simply involves feeding at different times. Two different species of fish that eat the same food can share the supply if one feeds at night and the other during the day. In the daytime, the reef teems with animals that are grazing, stalking, sifting, or chasing down their next meals. Near dusk, the daytime feeders retreat to find places to hide, and the nighttime feeders emerge. As one shift ends and the next one begins, there is a small lull in reef activity; however, by the time it is dark, the night feeders are hard at work.

Another way populations of reef fish share the bounty is by eating in different parts of the habitat. The seafloor is one of the reef habitats that fish subdivide at feeding time. Some bottom feeders dine on organisms living just below the sand and sediment. These predators float above the reef floor, watching it closely for any signs of activity. If something moves, they rush forward and grab it. Other fish on the bottom search for their prey by "mining," blowing streams of water over the sediment, or by stirring it to expose prey. A few species of bottom feeders have barbels, long, whiskerlike structures used to feel around in the sediment for prey. A different group of fish actually eats the sand and sediment, filters out the food, then expels the soil back into the environment.

A few species of fish have found niches for themselves by feeding on organisms that nothing else wants to eat. Some of these dine on sponges, despite the fact that sponges produce repulsive chemicals and their bodies contain sharp spicules. Other fish survive by attacking and eating long-spined black urchins, poisonous creatures that are covered with needle-sharp spines. Most animals give these urchins wide berths, but the highly specialized urchin eaters have developed methods of flipping the animals onto their backs, enabling them to attack the urchins' more vulnerable undersides.

Of the more than 22,000 fish species known worldwide, nearly one-third of them have been seen on coral reefs. There are two major groups of fish, the cartilaginous families, which include the skates and rays, and the bony fish families. Bony fish that are characteristic of coral reefs include damselfish and anemone fish (family Pomacentridae), squirrelfish and soldierfish (family Holocentridae), wrasses (family Labridae), parrot fish (family Scaridae), surgeonfish (family Acanthuridae), butterfly fish (family Chaetodontidae), angelfishes (family Pomacanthidae), groupers and sea bass (family Serranidae), blennies (family Blenniidae), gobies (family Gobiidae), cardinal fish (family Apogonidae), and grunts (family Haemulidae).

Sharks and Rays

Fossil evidence shows that early fish, the ancestors of today's bony fish, had skeletons made of cartilage. Today, the only fish that still have cartilaginous skeletons are the skates and rays. These two groups of primitive fish are closely related.

Several species of sharks and rays are attracted to the bounty of food that the reef has to offer. Although these types of fish have reputations as dangerous creatures, most are fairly shy and not aggressive toward humans. Like other reef residents, skates and rays spend their days looking for food. Because they have few predators, these two groups of animals do not invest much energy in defensive behaviors.

Shark Anatomy

Although there are many kinds of sharks, they all are similar anatomically. A shark's digestive system begins at the mouth, which is filled with teeth. Shark teeth are continuously produced, and at any time a shark may have 3,000 teeth arranged in six to 20 rows. As older teeth are lost from the front rows, younger ones move forward and replace them. Teeth are adapted to specific kinds of food. Depending on their species, sharks may have thin, daggerlike teeth for holding prey; serrated, wedge-shaped teeth for cutting and tearing; or small, conical teeth that can crush animals in shells.

The internal skeletons of sharks are made of cartilage, a lightweight and flexible bonelike material. Their external surfaces are very tough and rugged. Sharks have extremely flexible skin that is covered with placoid scales, each of which is pointed and has a rough edge on it. Shark fins are rigid and cannot be folded down like the fins of bony fish.

Like other aquatic organisms, sharks get the oxygen they need to live from the water. Compared to air, water contains a small percentage of dissolved oxygen. Surface waters may contain five milliliters of oxygen per liter of water, dramatically less than the 210 ml of oxygen per liter of air that is available to land animals. To survive, fish must be very efficient at removing and concentrating the oxygen in water.

In aquatic organisms, gills carry out the function of lungs in terrestrial animals. To respire, sharks pull water in through their mouths and *spiracles,* holes on top of their

Sharks on the Coral Reef

Zebra sharks (*Stegostoma fasciatum*) have black and white stripes when they are young but lose the stripes and develop spots as adults. These solitary night hunters have very flexible bodies and can swim into tight places to search the cracks and crevices of the reef for small fish, snails, and clams and other mollusks. During the day, they rest on sandy ocean floor near the reef, lying with their mouths open and facing the currents to keep water flowing over their gills. The adults reach lengths of 10 feet (3.1 m).

An easily identifiable reef animal is the nurse shark (*Ginglymostoma cirratum*), a species that is distinguished by skin flaps on the nose and barbels on the chin. This docile,

heads. The water passes over their gills and exits through the gill slits on the sides of the head. Most species of sharks can pump water over their gills by opening and closing their mouths. Some sharks, the "ram ventilators," must swim continuously to move water over their gills. Oxygen in water is picked up by tiny blood vessels in the gills, then carried to the heart, a small two-chambered, S-shaped tube. From there, oxygenated blood is pumped to the rest of the body.

Sharks fertilize their eggs internally. Males transfer sperm to females using modified pelvic fins. Some species are *oviparous,* which means the female lays fertilized eggs. Shark eggs may be deposited in lagoons or shallow reef water, where they incubate for six to 15 months. Many of the eggs' cases are equipped with hairy or leathery tendrils that help hold them to rocks or plants. Other species are *viviparous,* so the embryos develop inside the mother and are born alive. Several species are *ovoviviparous,* which means that the embryo develops inside an egg within the female's body. The egg hatches inside the mother, the hatchling eats the yolk and any unfertilized eggs, then is born alive.

Shark populations are relatively small compared to other kinds of fish. One reason is because shark reproduction rates are low. Unlike fish and many of the invertebrates, a female shark produces only a few offspring each year. In addition, the gestation period, time when the embryo develops inside the mother, of viviparous species is long.

Shark Senses

Humans have five senses—sight, smell, taste, touch, and hearing—that help them gather and interpret information about their environment. Sharks have these senses as well as others (see Figure 5.1). Of their five basic senses, smell plays the largest role, and hearing plays the smallest. Sharks smell by detecting molecules in the water, similar to the way air-breathing vertebrates detect odor molecules. The sense of smell in sharks is so keen that they can distinguish one drop of blood in 25 gallons (115 l) of water. Some species have sensory barbels near their mouths that can pick up tastes in the seawater.

After a shark senses prey, it homes in by traveling up the prey's "smell corridor," moving side to side to read clues in the water. As the shark gets closer, the corridor of clues narrows. Once the prey is found, the shark grabs it, unhinging its jaw if necessary to get its mouth around large animals. Since prey thrash around and can injure their attackers, some sharks have special nictitating membranes, thick films that cover their eyes during the final moments of the attack.

Lateral lines are sense organs located along the sides of the shark body that can detect vibrations in water. Short tubes connect the lateral lines, which are made of fluid-filled canals just under the skin, to external pores. When vibrations in water strike the pores, the lateral lines detect them, providing sharks with information about their source. Vibrations could indicate anything from a school of fish to a wounded animal.

Shark snouts possess receptors called ampullae of Lorenzini that can sense electrical energy. The cells of the receptors are sensitive enough to pick up the very small electrical impulses produced by nerves of living things. The receptors work with the nervous system to help sharks find prey that are buried in sediment or wandering out of sight in the open sea. Like all of the shark's senses, the ampullae of Lorenzini are supported by, and themselves support, the other sensory organs. Input from all of the shark's senses give the animal a clear picture of everything in its environment.

unaggressive bottom feeder rests on ledges and in caves. When feeding, a nurse sharks sucks up its food by creating a vacuum that pulls the prey into its mouth. Some of this shark's favorite foods are mollusks and crustaceans, which it crushes with rounded teeth. Adult nurse sharks can reach lengths of 8 feet (2.4 m).

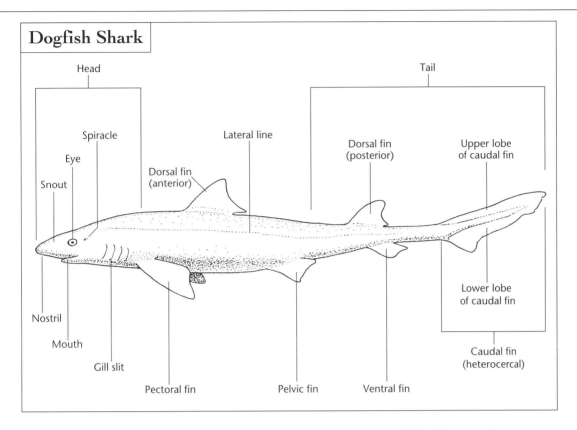

Dogfish Shark

Head

Tail

Spiracle

Lateral line

Dorsal fin
(posterior)

Upper lobe
of caudal fin

Eye

Snout

Dorsal fin
(anterior)

Lower lobe
of caudal fin

Nostril

Mouth

Gill slit

Pectoral fin

Pelvic fin

Ventral fin

Caudal fin
(heterocercal)

Fig. 5.1 Typical shark external anatomy is displayed by the dogfish shark. Special sensory structures include the lateral line and the ampullae of Lorenzini, located inside the snout.

The blacktip reef shark (*Carcharhinus melanopterus*) has a white flash on its side and a black tip on its tail. Active swimmers and aggressive hunters, blacktips work together to round up schools of fish into compact groups before attacking in a feeding frenzy. These sharks grow to be about 5 feet (1.5 m) long. In contrast, the whitetip reef shark (*Triaenodon obesus*) is very docile, spending the day lying beside other whitetips in a cave or on a ledge. These animals feed at night, moving around the reef shyly, hunting octopuses, lobsters, crabs, and small fish on the floor of the reef. The 5-foot (1.5-m) long animal can easily be identified by the white markings on the tips of their fins.

A Japanese wobbegong shark (*Orectolobus japonicus*) may be difficult to spot on the reef. This quiet animal has splotchy skin that camouflages it against the reef's soil and rocks. The wobbegong lies patiently on the bottom, grabbing prey that unwittingly swim too close. The shark grasps its chosen meal in daggerlike teeth, letting it squirm till exhausted. Then the wobbegong turns the prey so that it can be swallowed head-first, a technique that keeps the victim's fins from getting caught in the shark's throat.

The reef's smooth hammerhead shark (*Sphyrna zygaena*) is one of 10 species of hammerheads, all of which have flattened projections at the sides of their heads. The shark's eyes are mounted on the outer edges of the head lobes, and their nostrils are set far apart. Hammerheads can grow to lengths of 14 feet (4.3 m), and they aggressively hunt fish and rays, although they will scavenge, too. In the summer, groups of hammerheads may migrate to cooler water.

Skates and Rays

Nicknamed the "pancake sharks," skates and rays have a lot in common with their cartilaginous cousins; however, their outward appearances are quite different from those of sharks. The bodies of skates and rays, such as the blue-spotted stingray in Figure 5.2, are broad and flattened instead of torpedo-shaped, as in sharks, and the gill slits of skates and rays are located under the body. Skates and rays feed on shallow, sandy reef flats where they can pin down their prey with their winglike fins. When not feeding, they often lie on the bottom to rest. For protection, many species possess one or more poisonous spines on the upper surface of their tails.

The southern stingray (*Dasyatis americana*) is usually inactive during the day but hunts most of the night by moving over the bottom of the reef. To uncover prey, it spews jets of

water from its mouth into the sediment, or stirs up the sand and silt by vigorously flapping its pectoral fins. Once food is located, the ray lifts its body off the bottom with a pectoral disk and moves the prey into its mouth.

The male southern stingray is much smaller than the female, maturing at a disk width of about 1.5 feet (0.46 m). Females average about 3 feet (0.91 m) wide, but may grow to 6 feet (1.8 m). After mating, the embryos are carried in the mother and nourished first by yolk and then by special "uterine milk" that is produced in cells that line the stingray's uterus. After about five months, three to five young are born.

Venomous barbs on the tails of southern stingrays can be used for protection if the animals are disturbed; however, their venom pales in comparison to the venom of the blue-spotted stingray (*Taeniura lymma*). This yellow fish is covered with bright blue dots to advertise its dangerous nature; the toxin is strong enough to injure, or even kill, humans.

Fig. 5.2   The blue-spotted stingray hunts for food on the reef floor. (Courtesy of Getty Images)

Colorization

One of the most striking features of fish is their colorization. Coloring and body marks on fish help them avoid predators by staying out of sight. Many prey species, such as gulf flounder, avoid being eaten by blending in with their surroundings, matching the subtle shades of their habitats. Spotted fish look like the seafloor, and striped fish blend in with grasses. Some reef fish display bright colors because they live among brightly colored sponges and corals.

Conversely, coloring that mimics fish's habitats helps predators get close to their prey. The ability to avoid detection is a significant advantage for such hunters as scorpion fish that wait quietly until prey comes within striking distance. A hunter is able to conserve both time and energy if it does not have to pursue its food.

Most fish, including sea trout and grouper, display some degree of countershading. This form of coloring reduces the clarity of the fish's body outline in water. The simplest, and most common, form of countershading is a dark dorsal side and a pale ventral side, with intermediate colors between the two. When sunlight filters down through the water, it lightens the fish's back and throws shadows on its underside. The overall effect of countershading lessens the degree of contrast between the fish and the water.

A few species of fish, such as the spotfin butterfly fish and the high hat fish, show disruptive or deflective colorization that includes bands, stripes, or dots of contrasting colors. These colors and patterns confuse predators by distorting the true shape, size, and position of the fish. Bright patterns draw the predator's eye, causing it to see the pattern rather than the fish itself. This type of coloring can deflect the predator's attention away from a fish's vulnerable areas, such as its head and eyes.

Colorization can also be used as an advertisement. There is no point in being poisonous and unpalatable if no one knows it. Instead of hiding, poisonous fish announce their dangerous status. Fish may also advertise their age or sex with coloring. Males are generally more colorful than females, whose duller shades help camouflage and protect them. Young fish may be transparent or pale, making it hard for predators to spot them, as well as letting the older fish of their own species know that they are not a threat.

Bony Fish

More than 90 percent of the world's fish have skeletons made of bone instead of cartilage. Bony fish appear in all sizes, shapes, and colors imaginable. They can be as small as guppies, or as massive as an 880-pound (400-kg) tuna. Some are shaped like bullets, some like pencils, and others like flat plates. On the reef, bony fish display a rainbow of colors and an almost endless variety of markings.

When people envision a coral reef fish, most think of parrot fish. These large, brightly colored animals have teeth that are fused to form a parrotlike beak, giving them their name. They are usually found feeding in shallow reef water where they scrape off alga that is growing on top of the coral skeletons, or chew up the skeletons to get to the algae within. Since skeletal material is extremely hard, the digestive systems of parrot fish have developed unique adaptations to handle it. Teeth in their throats grind the calcium carbonate and release the algae. The algae travel on through the digestive system where they are broken down, absorbed, and used for nutrition. The pulverized coral is defecated as sand.

Because they physically remove reef material, parrot fish play important roles in sculpting the reef structure. In one year, a parrot fish can convert about five tons of reef into sand. Their activity is a major part of the bioerosion of the reef and part of the destructive forces that reduces its size. At the same time, the activities of parrot fish help keep the coral animals alive. As grazers, they prevent large mats of algae from covering the coral and smothering it. In this way, they help build up the reef structure.

Parrot fish have developed a unique strategy to protect themselves from predators. At night, they find a crack or crevice in the reef that they can use as a "bedroom." Once inside, the fish spend about 30 minutes secreting sticky mucus which they wrap around their bodies, similar to a cocoon. The mucus seals in the odors of the parrot fish and prevents nighttime predators from finding them by smell.

One of the many species of parrot fish is the queen parrot fish (*Scarus vetula*) that grows to about 2 feet (60 cm). Like some other species of parrot fish, this one has two color patterns that are associated with its unusual sexual development. All of the young adult parrot fish are drably colored females, but some of them change into colorful blue and yellow males. A few develop into supermales, very brightly colored fish that have the first opportunity to spawn with the females and therefore the best chance to pass their genes to the next generation.

Bony Fish Anatomy

All bony fish share many physical characteristics, which are labeled in Figure 5.3. One of their distinguishing features is scaly skin. Scales on fish overlap one another, much like shingles on a roof, protecting the skin from damage and slowing the movement of water into or out of the fish's body.

Bony fish are outfitted with fins that facilitate maneuvering and positioning in the water. The fins, which are made of thin membranes supported by stiff pieces of cartilage, can be folded down or held upright. Fins are named for their location: Dorsal fins are on the back, a caudal fin is at the tail, and an anal fin is on the ventral side. Two sets of lateral fins are located on the sides of the fish, the pectoral fins are toward the head, and the pelvic fins are near the tail. The caudal fin moves the fish forward in the water, and the others help change direction and maintain balance.

Although fish dine on a wide assortment of food, most species are predators whose mouths contain small teeth for grasping prey. Nutrients from digested food are distributed through the body by a system of closed blood vessels. The circulation of blood is powered by a muscular two-chambered heart. Blood entering the heart is depleted of oxygen and filled with carbon dioxide, a waste product of metabolism. Blood collects in the upper chamber, the atrium, before it is pushed into the ventricle. From the ventricle, it travels to the gills where it picks up oxygen and gets rid of its carbon dioxide. Water exits through a single gill slit on the side of the head. The gill slits of fish are covered with a protective flap, the operculum.

In many bony fish, some gases in the blood are channeled into another organ, the swim bladder. This organ is essentially a gas bag that helps the fish control its depth by adjusting its buoyancy. A fish can float higher in the water by increasing the volume of gas in the swim bladder. To sink, the fish reduces the amount of gas in the bladder.

Most bony fish reproduce externally. Females lay hundreds of eggs in the water, then males swim by and release milt, a fluid containing sperm, on the eggs. Fertilization occurs in the open water, and the parents swim away, leaving the eggs unprotected. Not all of the eggs are fertilized, and many that are fertilized will become victims of predators, so only a small percentage of eggs hatch.

Fig. 5.3 The special features of bony fish include bony scales (a), opercula (b), highly maneuverable fins (c), a tail with its upper and lower lobes usually of equal size (d), a swim bladder that adjusts the fish's buoyancy (e), nostrils (f), pectoral fins (g), a pelvic fin (h), an anal fin (i), lateral lines (j), dorsal fins (k), and a stomach (l).

Features of Bony Fish

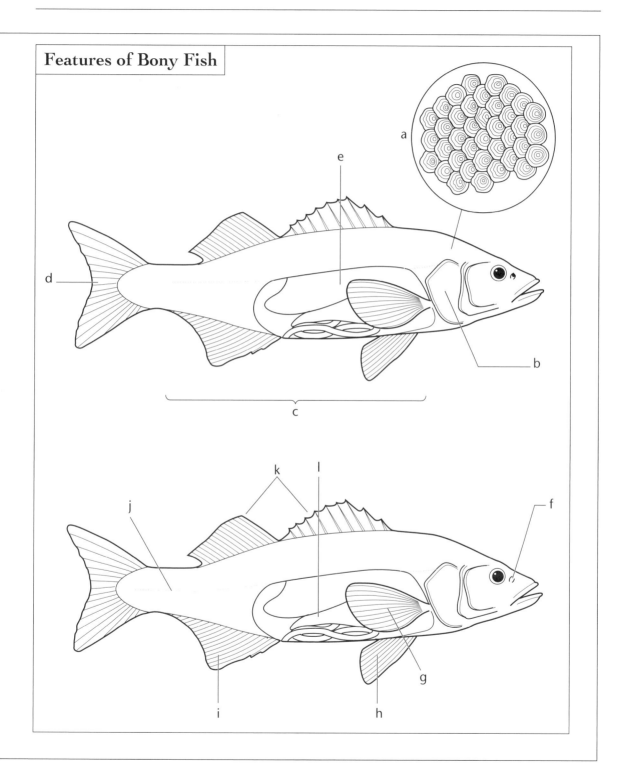

Damselfish, Clown Fish, Cardinal Fish, and Squirrelfish

Another colorful and abundant reef fish is the damselfish, a delicate-looking animal with a small mouth. Damselfish nibble on the encrusting algae that grow on coral and rock. These red, yellow, orange, or blue fish grow to be about 6 inches (15 cm) long.

Like most damselfish, the Pacific gregory (*Stegastes fasciolatus*) is active during the day. Most of this fish's time is spent "farming" a patch of alga that may cover up to 10.7 square feet (1 sq. m) of the reef. An excellent gardener, the little damselfish weeds out unwanted corals and algae so that its favored algae can flourish. To supplement plant growth, the fish fertilizes its garden with its own feces.

One of the brightest and best-known reef fish is the percula clown fish (*Amphiprion ocellaris*), one of several species of clown fishes and a member of the damselfish family. Reaching only 2.4 inches (6 cm), this small animal is striped in orange, white, and black. The percula clown fish lives in the tentacles of several kinds of anemones, despite the latter's deadly stinging cells. Anemone tentacles are covered with mucus that prevents one tentacle from stinging another. The clown fish coats its body with this same mucus, a process that takes about one hour. If the clown fish and anemone are separated, the fish loses its immunity and has to repeat the procedure.

From its safe haven among the tentacles, a clown fish can cautiously dart out to feed on plankton and small crustaceans. The female percula clown fish lays her eggs on rock or coral near the anemone. After the male fertilizes them, both parents guard the clutch until the eggs hatch, running back to their safe house when threatened.

Squirrelfish, such as those in the lower color insert on page C-6, are colorful animals whose large populations make them conspicuous on many reefs. Sabre squirrelfish (*Sargocentron spiniferum*) are one of the larger species, growing to lengths of 17.7 inches (45 cm). Squirrelfish are numerous, but their populations are rivaled by the cardinal fish. During the day, cardinals hide in the crevices, but at night these fish, which average about 4 inches (10.2 cm) long, cover the reef. Many are red, a color that looks black in dark water and helps them avoid predators. Like other fish that are active at night, cardinal fish have very big eyes.

During the reproductive cycle, cardinal fish employ a very unusual technique of guarding their eggs called mouth brooding. After eggs are laid in the water and fertilized, the male takes them in his mouth and keeps them there until they hatch. During this period of incubation, the male fish does not eat. Even after hatching, the young may return to hide in the father's mouth for short periods of time.

The cardinal fish is just one of several kinds of fish that are difficult to spot in the water. Many reef fish are well camouflaged, an adaptation that offers two big advantages: It helps animals avoid their predators, and it makes it easier for predators to ambush their own prey. The longlure frogfish (*Antennarius multiocellatus*) is so highly camouflaged that it looks like one of the corals or sponges. When the longlure frogfish moves from one location to another, it changes its colors so that wherever it goes it blends in perfectly. To attract prey, the fish is equipped with a fleshy antenna that dangles in front of its head similar to a fishing lure dangling from a pole. By waiting quietly with its antenna poised, the frogfish attracts small fish that mistake the lure for a snack. If a fish happens to bite off the lure, it grows back.

Scorpion Fish, Catfish, and Eels

Another fish that depends on camouflage is the scorpion fish (family Scorpaenidae), a heavy animal with ridges and spines on its back. The fins and spines contain venom that protects this fish from predators. Warts and skin tassels covering its body make the scorpion fish look like part of the reef. This animal waits patiently for prey to get close then literally sucks it in.

The bandtail puffer (*Spheroides spengleri*), like other species of puffers, has a very different way of protecting itself. By filling its abdomen with water, the little fish can inflate its body, giving it a bigger and fiercer appearance. In addition, an inflated fish is more difficult for a predator to swallow than an uninflated one.

Schooling

Fish often school, or swim together, in large groups. Schools of fish may be polarized, with all the fish swimming in the same direction, or non-polarized, with fish swimming in many directions. Both types of schools improve an individual fish's chance of survival. A large school of fish may be able to confuse a predator into thinking that it is one big, dangerous organism instead of a group of small, helpless fish. In addition, if a fish is in a school, it stands a good chance of being spared when a predator does attack. Plus, a school of fish has more lookouts than one fish swimming alone, and is more likely to notice danger.

Some kinds of fish, such as groupers, only form schools when it is time to *spawn*. This strategy ensures that the males and females will release their gametes into the water at the same time. If egg predators are nearby, they will eat some of the eggs but may not be able to eat all of them, so some will probably survive. Foraging for food can also bring a group of fish together. As a school, foraging fish like mackerel and herring have plenty of sets of eyes that improve the chances of finding something to eat. By working as a team, the school may be able to overwhelm and take prey that one fish alone could not handle.

A close relative, the porcupine fish (*Diodon hystix*), uses the same mechanism when it feels threatened. Porcupine fish are covered with damaging sharp spines that deter predators. Shallow reef waters and turtle grass beds are some of the places where porcupine fish hunt for prey. Both porcupine fish and bandtail puffers have two fused teeth in each jaw that give their mouths a beaklike appearance. This type of jaw is efficient at crushing the hard shells of crabs and mollusks.

The moray eel (family Muraenidae), seen in the upper color insert on page C-7 has a fearsome reputation because its gaping mouth is full of sharp teeth. In reality members of this group are shy and retiring animals that spend their days hiding in reef crevices. Morays hold their mouths open so that water can be pumped over the gills. Their long, muscular bodies are highly modified for living among rocks and coral. Morays are scale-less, and both their pectoral and pelvic fins are absent. In addition, their remaining fins are fleshy ridges that are covered with thick skin. This body streamlining is an adaptation that helps morays navigate through the narrow passages of the reef structure. When they emerge at night to look for invertebrates, they swim by moving their entire bodies back and forth in an S-shaped pattern.

Another long, fierce-looking reef fish is the great barracuda (*Sphyraena barracuda*), a slender animal whose body ends in a pointed head with a narrow snout and sharp teeth. Barracudas usually hover around the edges of reefs, waiting for prey to appear. Once a potential meal is spotted, they rush forward aggressively to grab it. Some barracudas reach impressive lengths of 6 feet (1.8 m).

The only catfish on the reef are the coral catfish (*Phetusus angularis*). These small fish have venomous notched spines near their dorsal and pectoral fins. When young, coral catfish school, or live in groups, by the hundreds near the reef floor; however, as they get older, they form smaller groups of about 20 animals. Maturity also changes they way they look. Young catfish are black, but adults develop a brown color with yellow or white horizontal stripes. They feed on crustaceans, mollusks, and worms, which they find by stirring up the reef floor.

Grunts, Wrasses, Gobies, and Flounders

A reef is a pretty noisy place to live. One family of fish, the grunts, get their name from the grunting noise they make by grinding their pharyngeal teeth. Grunts are pretty, deep-bodied fish that are generally found traveling in schools. The French grunt (*Haemulon flavolineatum*) has yellow stripes on a silver background or blue stripes on orange.

Wrasses are a large group of fish that are well represented on the reef. Most are small, averaging about 9 inches (23 cm) long, with cigar-shaped bodies, pointed snouts, and prominent canine teeth. During the daytime wrasses feed, and at night they hide in crevices or buried in sand.

Wrasses feed in several ways, depending on their species. The cleaner wrasses remove mucus, parasites, and scales from the bodies of larger fishes. Another species exhibits "following behavior" to find prey. As larger fish swim through the substrate and disturb it, these wrasses follow closely, picking up invertebrates that the large fish overlook. A different species of wrasses searches the reef for invertebrates too small for most fishes to prey upon. Finally, some wrasses flip over rocks and pieces of coral with their snouts, looking for hidden invertebrates.

There are more species of gobies on the reef than any other kind of fish. These small animals, which measure less than 2 inches (5 cm), have prominent eyes set high on their heads.

Gobies spend most of their time hiding among rocks, corals, or shells. Some make their homes inside of other living things like sponges and the shells of mollusks, and a few share burrows with shrimp.

The peacock flounder (*Bothus lunatus*) is a reef flatfish that usually hides in the sand and sediment on the reef floor. Like other kinds of fish, the peacock flounder begins life with bilateral eyes and a mouth in the middle of its face; however, as the fish matures, one eye begins to migrate toward the other one so that by maturity both eyes and the mouth are located on one side of the head. Peacock flounders do not have swim bladders, so they sink to reef floor and remain there, often wriggling around enough to bury their bodies in the sand with just their eyes and gill opercula exposed. The skin of a peacock flounder is covered with brown spots, but can change colors to camouflage the fish.

Sea Horses, Surgeonfish, and Remoras

Sea horses and their close relatives, the pipefishes and sea dragons, have a distinctive appearance. Unlike other bony fish, the elongated bodies of these animals are supported by rings of bone. All of these fish feed through tubelike mouths that suck in small crustaceans and other prey. Instead of chasing their food, they quietly hover in the water and wait for something edible to swim by.

A sea horse, shown in Figure 5.4, swims upright among the corals or sea grasses, moving slowly from one perch to another. To keep from drifting away, a seahorse may hold in place by wrapping its prehensile tail around a plant or piece of coral. Pipefish and sea dragons swim like other kinds of fish, with their heads leading. A pipefish has a long, thin shape, so it resemble grass or reeds. In contrast, sea dragons are covered in leaflike appendages that help camouflage them as they float in beds of leafy seaweeds.

In all groups, the males take care of the eggs. After eggs form in the female's body, they are transferred to a thin-skinned pouch on the male. He fertilizes the eggs, then carries them until they develop and hatch, providing them protection from predators.

The surgeonfish, a group that includes the tangs, are represented by more than 70 species on the reef. All are brightly colored and have oval bodies that are wide and flat. Surgeonfish get their name from the scalpel-like scales that grow at the base of their tails. These razor sharp scales remain folded neatly against the body most of the time but become erect when they sense danger. Surgeonfish can seriously wound predators by swimming beside them, lashing their bodies back and forth.

Most surgeonfish species are herbivores. They swim around the reef in small groups grazing algae, very much like a herd of cows might roam through a pasture. Their small incisorlike teeth are adept at scraping algae off the rocks and reef. Occasionally, a large group of surgeonfish will overwhelm a territorial damselfish and gorge on the "farmer's" algal garden.

A close relative of the surgeonfish is the graceful Moorish idol (*Zanclus cornutus*), which can grow to 8 inches (20 cm) and is distinguished by a long dorsal fin. Using its protruding lips and long snout, the Moorish idol probes into crevices for food that many other fish are unable to reach. Sponges are the fish's favorite food, but it also feeds on algae.

Butterfly fish (family Chaetodontidae) are some of the most colorful reef fish. These beautiful animals flitter around the reef with a butterfly-like motion. They primarily feed on coral polyps, although they may also consume small invertebrates. When feeding, butterfly fish tend to bite off several polyps within a small area, then move on to another place, a technique that does not kill the coral. Most live in male-female pairs, and together they claim and protect regions of coral from intruders.

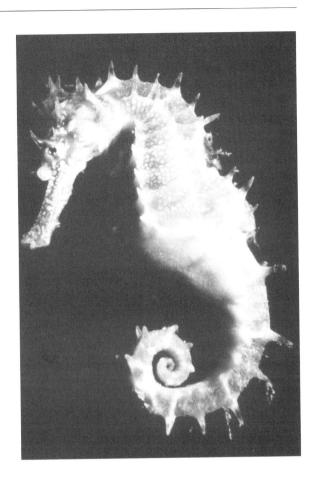

Fig. 5.4   A sea horse clings to marine plants with its prehensile tail.
(Courtesy of Getty Images)

Territoriality

Fish living close together may show territorial behavior, the tendency to occupy and defend an area, usually to eat and reproduce there. There are many different patterns of territorial behavior. Some species are territorial all the time, but others may only display this behavior during reproductive periods. Depending on the situation, fish may be territorial against their own species, toward other species, or both.

Territorial behavior requires a lot of energy, and a fish cannot afford to expend more energy on defending its territory than it takes in as food. For this reason, fish have developed several threatening displays that involve a lot of posturing and ritualized motions yet conserve energy. If an intruder gets close to a damselfish's alga garden, for example, the damselfish first attempts to scare it away with a threatening posture of spread fins and gill covers. If this strategy does not solve the problem, the damselfish makes excited, aggressive movements. Only as a last resort will it attempt to chase away an intruder, as actual chase could lead to a fight that might end in the death of the defender.

The flattened bodies of butterfly fish offer them some protection from predators. They can easily slip into cracks in the reef where they open their fins to wedge their bodies tightly in place. The fish are so thin that they seem to disappear when they face a predator head on. Butterfly fish, such as the spotfin butterfly fish in the lower color insert on page C-7, further confuse their enemies with a large eyespot near their tails. These marking can fool a predator into lunging for the wrong end of the animal.

There are many different species of angelfish. Food choices vary by species, but as a group angelfish are daytime feeders that favor sponges, tunicates, and algae, nibbling at them with their brushlike teeth. Most species are very territorial, often occupying and defending a small cave or some other space for several years.

As in parrot fish, all individuals in certain species of angelfish begin life as females. As they mature, a number of the females change into males. Color plays an important role in angelfish interactions. The two sexes can be visually distinguished by their slightly different coloring, and the marking and colors of juveniles are different from those of adults. This may protect the young fish from aggressive adults of their own species.

Remoras are slim fish that are usually 1 to 3 feet (0.9 m) long. Their dorsal fins are modified into sucking disks, which they use to attach themselves to sharks, whales, rays, or turtles. This behavior may benefit the remora, who often eats the leftovers from its host's meals. Being attached to a big animal also protects the smaller fish from predators. If it chooses to do so, the smaller fish can swim away from its host to forage. If no hosts are available, remoras will form schools that swim around in spiral patterns, with the largest animals on bottom and the smallest on top.

Conclusion

Coral reef communities support a larger number and greater diversity of fish than any other aquatic habitat. Reef fish are specialized for a variety of feeding strategies and habitats, adaptations that permit different species to feed within the same area and on similar food supplies. Damselfish, for example, primarily feed on algae that grow on top of the coral skeletons, but parrot fish prefer the algae within the coral polyps.

Many reef fish are predators of fish or invertebrates. Even though the most impressive hunting fish are large, there are an equal number of smaller, less obvious ones. Hunters patrol the area day and night, even at dusk and dawn when fish are moving to their refuges.

A good number of reef fish are plankton eaters, including soldierfish and cardinals. Plankton feeders are not closely related to one another but share similar lifestyles. Usually they have small, streamlined bodies and forked tails, which are good at delivering bursts of speed to escape predators. They tend to feed in groups for safety.

The plant-eating reef fish include surgeonfish and damselfish, animals that use their sharp-edged teeth to clean the coral reef of algae. Alga has a tendency to grow quickly and can eventually smother and "choke" a reef to death if it is not kept in check. Grazing fish are critical to maintaining the balance of algal cover on the reef while at the same time supplying themselves with a source of energy.

Large populations and intense competition have led to the development of reef fish that fill every niche of the ecosystem. Scientists are still working to discover all of the fish that make their homes around the reef and to understand the evolutionary pressures that have resulted in these varied and colorful reef residents.

6

Reptiles, Birds, and Mammals
The Top of the Coral Reef Food Chain

*I*n the ecosystems of coral reefs, vertebrates, or animals with backbones, are the dominant organisms. Four of the five vertebrate groups—amphibians, fish, reptiles, birds, and mammals—are represented there. Fish are the largest group, followed by birds, mammals, and reptiles. Amphibians, land-based animals with soft skin and lungs, are absent.

Unlike the majority of the invertebrates, numerous vertebrates are the top consumers in the reef food chains. In this position, they have few, if any predators, so they are free to spend their time feeding, hunting, participating in courtship behaviors, and raising young. These animals live in spaces such as the edges of the reef or along the surface of the water, interacting with only some of the other reef residents.

Marine Reptiles

Among the reptiles, only a small number of species have become adapted for life in the sea. Of these, just a few are full-time reef residents. Reptiles frequently found around reefs include the sea turtles and sea snakes.

Sea turtles move through the water with the grace of ballerinas. These large reptiles, related to the more familiar but generally smaller land turtles, are superbly adapted for sea life. Their limbs are modified as strong flippers that effortlessly push their streamlined bodies through the water.

Turtles are such expert swimmers and so perfectly designed for their lifestyles that it would be easy to think of them as big fish; however, unlike fish, turtles must swim to the surface to breathe. These animals have lungs instead of gills, and they breathe through their noses. Even though active turtles need to surface every few minutes, they are capable of staying under water for long periods of time by holding their breath. When resting in caves or on ledges, sea turtles may remain submerged up to two and one-half hours.

As true marine animals, sea turtles rarely leave the ocean. The only reason they visit the shore is to lay eggs, and then only the females emerge. Nesting females always return to the beaches where they were born to lay their eggs (see Figure 6.1). Males accompany them only as far as the shallow water.

Several species of sea turtles occasionally visit the coral reef; however, two species are reef residents: the Atlantic hawksbill sea turtle (*Eretmochelys imbricata*) and the green sea turtle (*Chelonia mydas*). From time to time they may be joined by other species of marine turtles that take time out of their migratory travels to rest and eat in luxurious reef accommodations.

The green sea turtle is an impressive animal that grows up to 3.5 feet (1.1m) long and can weigh 400 pounds (181 kg). As seen in the upper color insert on page C-8, its carapace,

Fig. 6.1 Sea turtles dig nests on sandy beaches where they lay their eggs.

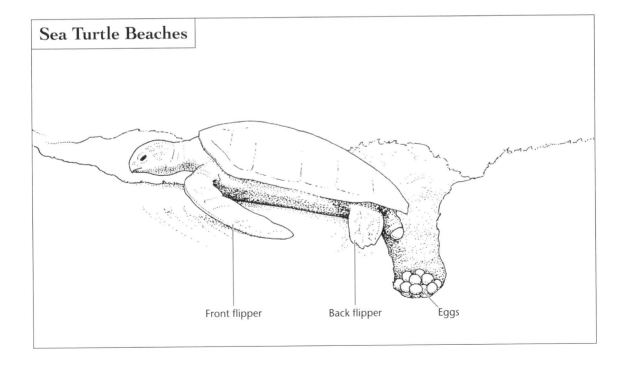

Sea Turtle Beaches

Front flipper   Back flipper   Eggs

Marine Reptile Anatomy

Reptiles are not usually associated with marine environments. In fact, of the 6,000 known species of reptiles, only about 1 percent inhabits the sea. Members of this select group include lizards, crocodiles, turtles, and snakes. Each of these organisms shares many of the same anatomical structures that are found in all reptiles: They are cold-blooded, air-breathing, scaled animals that reproduce by internal fertilization. Yet, to live in salt water, this subgroup has evolved some special adaptations not seen in terrestrial reptiles.

In turtles, the shell is the most unique feature. The lightweight, streamline shape of the shell forms a protective enclosure for the vital organs. The ribs and backbone of the turtle are securely attached to the inside of the shell. The upper part of the shell, the carapace, is covered with horny plates that connect to the shell's bottom, the plastron. Extending out from the protective shell are the marine turtle's legs, which have been modified into paddle-like flippers capable of propelling it at speeds of up to 35 miles per hour (56 kph) through the water. These same legs are cumbersome on land, making the animals slow and their movements awkward.

Most air-breathing vertebrates cannot drink salty water because it causes dehydration and kidney damage. Seawater contains sodium chloride and other salts in concentrations three times greater than blood and body fluids. Many marine reptiles drink seawater, so their bodies rely on special salt-secreting glands to handle the excess salt. To reduce the load of salt in body fluids, these glands produce and excrete fluid that is twice as salty as seawater. The glands work very quickly, processing and getting rid of salt about 10 times faster than kidneys. Salt glands are located on the head, often near the eyes.

There are more than 50 species of sea snakes that thrive in marine environments. Sea snakes possess adaptations such as nasal valves and close-fitting scales around the mouth that keep water out during diving. Flattened tails that look like small paddles

the upper shell, is mottled in shades of dark brown on top and creamy white below. This type of dark-on-top, light-on-the-bottom coloration, called countershading, makes turtles hard to see in the water. From above, their carapace looks like the seafloor, and from below the plastron, the lower shell, blends in with the sky. Such camouflaging helps turtles get close to their prey before striking. It also helps them avoid sharks, their only predators.

Every two or three years, sexually mature green sea turtles make long journeys to mate and lay their eggs. They leave

easily propel these reptiles through the water. The lungs in sea snakes are elongated, muscular air sacs that are able to store oxygen. In addition, sea snakes can take in oxygen through the skin. Their adaptations to the marine environment enable sea snakes to stay submerged from 30 minutes up to two hours; however, this ability comes at a cost. Because marine snakes routinely swim to the surface to breathe, they use more energy and have higher metabolic rates than land snakes. To balance their high energy consumption, they require more food than their terrestrial counterparts.

Finally, crocodiles usually occupy freshwater, but there are some species that live in brackish water (in between salt water and freshwater) and salt water. These animals have salivary glands that have been modified to excrete salt. Their tails are flattened for side-to-side swimming and their toes possess well-developed webs. Saltwater crocodiles are equipped with valves at the back of the throat that enable them to open their mouths and feed underwater without flooding their lungs.

their feeding grounds and swim 600 miles or more, returning to the beaches where the females were born. During the months of March and April, mating occurs in offshore waters, then the females go ashore to lay their eggs. Green sea turtles have the ability to retain viable sperm for months after mating. The eggs laid at one mating were fertilized much earlier.

No longer supported by the water's buoyancy, female green sea turtles drag themselves across the beach to sandy spots above the tide line. While ashore, they shed sticky tears that keep their eyes moist and free of sand. Using powerful hind legs, each female digs an egg chamber, a task that may take the entire night. When the chamber is finally finished, the female deposits 100–200 eggs, each about the size of a Ping-Pong ball, then covers the nest and returns to the sea.

Hatching begins after 60 days of incubation, usually early in July. Working together, the hatchlings scrape sand off the roof of the nest and pack it into the nest floor, a strategy that builds up the nest until it is almost even with the beach. Using moonlight reflected in the ocean water as a beacon, all of the hatchlings scramble from the nest one evening and race toward the water. Some are picked off by birds and crabs on the beach, and others are grabbed by fish waiting for them in the shallow water.

Those that survive strike out on their own, swimming non-stop for the next 36 to 48 hours. When they get out far enough, the baby turtles are picked up by currents and carried

Fig. 6.2   A hawksbill sea turtle swims around the coral reef. (Courtesy of NOAA, Coral Kingdom Collection)

into the open ocean. Green sea turtles remain at sea for several years, feeding on jellyfish and other invertebrates. When they are juveniles, they return to the reef areas where adults are living. There, young turtles join the colony, grazing on algae among corals and rocks.

The Atlantic hawksbill sea turtle, pictured in Figure 6.2, is another reef resident. With an orange, brown, or yellow carapace that measures up to 35.8 inches (91 cm) long, the hawksbill can weigh 100–150 pounds (40–60 kg). The head of the turtle is narrow, with two pairs of scales in front of the eyes and a hawklike jaw that accounts for its name. The shape of this turtle's mouth is ideal for reaching into the cracks and crevices of coral reef where it finds sponges, octopuses, and shrimps. The hawksbill also eats squid that are swimming in the water column. After eating, it rests on the ledges and caves of the reef.

Nest building and egg laying occur every two to three years, preceded by mating in areas of shallow water. Male hawksbill turtles can be distinguished from females by their long tails. A female nests two to four times during each egg-laying season, depositing about 160 eggs in each clutch. Renesting females will usually return to the same part of the beach, often building her second and third nests within sight of the first one. After 50 to 70 days of incubation, hatchlings climb out of the nest and make their way to the sea. Mortality rates are very high, but the ones who survive swim out to sea. Like the green sea turtles, they are not usually seen again until they return to the area as juveniles.

Sea snakes make up 86 percent of the marine reptiles, and they are primarily found in tropical waters. Some species, including the olive sea snake (*Aipysurus laevis*), live around reefs. With a stout, round body that averages about 3.9 feet (1.2 m) in length, the snake varies in color, ranging from dark brown to purplish brown on the dorsal side, fading to light brown on the ventral surface. The flattened tail is creamy white and has a brown ridge on the dorsal side.

The olive sea snake prefers reef waters that are 16.4 to 147.6 feet (5 to 45 m) deep where they can prey on fish, fish eggs, cuttlefish, and crabs yet surface quickly for air. During the day, the snake feeds by weaving among coral structures in search of animals that are at rest. When prey is located, the snake uses constriction to hold the victim while it injects venom with its fangs. The olive sea snake's venom contains enzymes that begin digesting the prey from the inside.

After courtship in the open water, olive sea snake mating takes place on the reef floor. Competition for mates is fierce and several males may vie for a single female. The young snakes, which are born alive, grow quickly, maturing in four to five years.

Even though not as numerous as the olive sea snake, the turtle-headed sea snake makes it home on many reefs. Preferring lagoons to energetic parts of the reef, this snake is often found in large aggregates. The turtle-headed sea snake is a daytime feeder that slowly moves through the reef, seeking out small fish and crustaceans that it immobilizes with venom. It also feeds on the eggs of fish, such as gobies and blennies, that spawn in the lagoon, scooping them up from the reef floor with its hard pointed snout, the feature that most resembles a turtle's head. The snake is a voracious eater and feeds every two or three hours.

Seabirds

Birds, vertebrates of the class Aves, are the second largest group of vertebrates on Earth, following fish. Unlike marine reptiles, seabirds do not actually live in the water; however, they depend on the sea for their food, and their bodies are highly specialized for aquatic life. Many of the seabirds who feed on animals in the coral reefs also spend some of their time in other marine areas. All of them go to shore during parts of their lives, and some migrate from one ocean area to another.

Seabirds are among the longest-lived birds in the world, many with life spans of 30 years or more. For these animals, reproducing is a serious investment of time and energy. Compared to terrestrial avians, seabirds produce fewer offspring and the young are slower to mature, taking an average of seven years. Many choose their mates for life, and the males and females work together to incubate the eggs.

There are many species of birds that nest near coral reefs and interact with the reef food webs. Some of the largest seabird families represented on the reef include frigate birds (family Fregatidae), tropic birds (Phaethontidae), petrels (Procellariidae), boobies (Sulidae), terns and noddies (Sternidae), and albatrosses (Diomedeidae).

Magnificent frigate birds (*Fregata magnificens*) are striking black birds with deeply forked tails and wingspans of about 7.5 feet (24.6 m). They are easily recognized by their gular sacs, red membranous pouches that the males inflate during courtship. Unlike other seabirds, magnificent frigate birds do not produce a lot of preening oil. These animals rarely float or paddle in the ocean, even to gather food, so their wings require little waterproofing. Instead, most of their meals are stolen from other birds using a highly effective technique of harassment, pestering the victim so much that the irritated bird regurgitates its meal. When the stomach contents are finally disgorged, the magnificent frigate bird deftly catches the mass, often before it hits the water, and whisks it away. If there are no other birds feeding in the area, magnificent frigate birds, who are capable fishers, revert back to the predator mode and catch their own fish or squid from the water.

Reefs also support white-tailed tropic birds (*Phaethon lepturus*). The adults of this species have wingspans of about 37 inches (94 cm) and can be identified by the long central streamers of tail feathers, black markings on the wings, and yellow bills. Generally feeding at twilight, the white-tailed tropic bird flies high over the ocean, then gracefully dives a distance of 50–70 feet (15–20 m) in pursuit of fish and squid. To lessen the impact of the dive, the birds have shock-absorbent air-filled pouches on their chests. In nesting season, the female lays one pink-and-brown egg on the bare ground or among rocks.

Marine Bird Anatomy

Birds are warm-blooded vertebrates that have feathers to insulate and protect their bodies. In most species of birds, feathers are also important adaptations for flying. As a general rule, birds devote a lot of time and energy to keeping their feathers waterproof in a process called preening. During preening, birds rub their feet, feathers, and beaks with oil produced by the preen gland near their tail.

The strong, lightweight bones of birds are especially adapted for flying. Many of the bones are fused, resulting in the rigid type of skeleton needed for flight. Although birds are not very good at tasting or smelling, their senses of hearing and sight are exceptional. They maintain a constant, relatively high body temperature and a rapid rate of metabolism. To efficiently pump blood around their bodies, they have a four-chambered heart.

Like marine reptiles, marine birds have glands that remove excess salt from their bodies. Although the structure and purpose of the salt gland is the same in all marine birds, its location varies by species. In most marine birds, salt accumulates in a gland near the nostrils and then oozes out of the bird's body through the nasal openings.

The term *seabird* is not scientific but is used to describe a wide range of birds whose lifestyles are associated with the ocean. Some seabirds never get further out into the ocean than the surf water. Many seabirds are equipped with adaptations of their bills, legs, and feet. Short, tweezerlike bills can probe for animals that are near the surface of the sand or mud, while long, slender bills reach animals that burrow deeply. For wading on wet soil, many seabirds have lobed feet, while those who walk through mud or shallow water have long legs and feet with wide toes.

Other marine birds are proficient swimmers and divers who have special adaptations for spending time in water. These include wide bodies that have good underwater stability, thick layers of body fat for buoyancy, and dense plumage for warmth. In swimmers, the legs are usually located near the posterior end of the body to allow for easy maneuvers, and the feet have webs or lobes between the toes.

All marine birds must come to the shore to breed and lay their eggs. Breeding grounds vary from rocky ledges to sandy beaches. More than 90 percent of marine birds are colonial and require the social stimulation of other birds to complete the breeding process. Incubation of the eggs varies from one species to the next, but as a general rule the length of incubation correlates to the size of the egg: Large eggs take longer to hatch than small ones do.

A reef bird that is capable of both diving and skimming for food is Audubon's shearwater (*Puffinus iherminieri*). Its dark head and brown upper body are set off by a white belly and throat. To feed, Audubon's shearwater flies close to the water, alternately flapping and gliding, picking up small crustaceans and fish larvae that swim near the surface. If it spots a fish or squid in deeper water, the bird dives after it. Like most shearwaters, it lays a single egg inside a hole in seaside cliffs.

The red-footed booby is the smallest member of the booby family, having a wingspan of 36 to 40 inches (91.4 to 01.6 cm). Its coloring is unusual among birds because individuals can vary from white to brown. The adults have torpedo-shaped bodies, long pointed wings, distinctive bright red feet, and conical blue bills. Red-footed bobbies feed by diving for fish and squid. During mating season, a pair builds a nest in the tops of trees and usually lays two eggs; however, they only hatch one of the eggs, perhaps because competition for food is keen among marine bird populations.

Sooty terns, members of the group of birds known as "sea swallows," have forked tails, long, pointed wings, and slender bills that curve downward. Adults display distinctive black-and-white plumage, but juveniles have sooty-colored feathers on their heads and chests. Favorite foods of sooty terns include small fish and crustaceans. Each spring, thousands of the birds migrate to tropical islands to form nesting colonies that cover acres of ground. Within these sprawling assemblies, new parents work together to create nurseries where all of the young birds are kept and protected.

Marine Mammals

Mammals are the most obvious group of animals on land, but they are relatively rare in marine environments. There are just a few types of mammals whose bodies have become specialized for marine life. Among these are the cetaceans, a group that includes whales and dolphins, porpoises, and dugongs.

Marine Mammal Anatomy

Mammals are warm-blooded vertebrates that have hair and breathe air. All females of this group have milk-producing mammary glands with which to feed their young. Mammals also have a diaphragm that pulls air into the lungs and a four-chambered heart for efficient circulation of blood. The teeth of mammals are specialized by size and shape for particular uses.

Marine mammals are subdivided into four categories: cetaceans, animals that spend their entire lives in the ocean; sirenians, herbivorous ocean mammals; pinnipeds, web-footed mammals; and marine otters. Animals in all four categories have the same characteristics as terrestrial mammals, as well as some special adaptations that enable them to survive in their watery environment.

The cetaceans, which include whales, dolphins, and porpoises, have streamlined bodies, horizontal tail flukes, and paddle-like flippers that enable them to move quickly through the water. Layers of blubber (subcutaneous fat) insulate their bodies and act as storage places for large quantities of energy. Their noses (blowholes) are located on the tops of their heads so air can be inhaled as soon as the organism surfaces above the water.

Manatees and dugongs are the only sirenians. These docile, slow-moving herbivores lack a dorsal fin or hind limbs but are equipped with front limbs that move at the elbow, as well as with a flattened tail. Their powerful tails propel them through the water, while the front limbs act as paddles for steering.

The pinnipeds—seals, sea lions, and walruses—are carnivores that have webbed feet. Although very awkward on land, the pinnipeds are agile and aggressive hunters in the water. This group of marine mammals is protected from the cold by hair and blubber. During deep-water dives, their bodies are able to restrict blood flow to vital organs and slow their heart rates to only a few beats a minute, strategies that reduce oxygen consumption. All pinnipeds come onto land or ice at breeding time.

The sea otters spend their entire lives at sea and only come ashore during storms. They are much smaller than the other marine mammals. Even though otters are very agile swimmers and divers, they are clumsy on shore. Their back feet, which are flipperlike and fully webbed, are larger than their front feet. Internally, their bodies are adapted to deal with the salt in seawater with enlarged kidneys that can eliminate the excess salt.

Spinner Dolphins

Spinner dolphins (*Stenella longirostris*) are easily spotted as they swim in the clear waters of coral reefs. Named for their ability to spin during acrobatic jumps, these dolphins are slender, with long thin beaks, sloping foreheads, and a stripe that runs from the eyes to the flippers. An adult measures 4.25 to 6.89 feet (1.3 to 2.1 m) long, and weighs between 100 and 165 pounds (45 and 75 kg). There are several varieties of spinners in different geological locations, and they vary slightly in shape and color.

Spinner dolphins feed at night on fish and squid in the deeper waters of the reef, although they will also eat organisms that live on the reef floor. Their mouths are equipped with 45 to 65 pairs of sharp teeth in each jaw. After feeding, spinner dolphins can often be found resting in protected areas of shallow water.

As social animals, spinners depend on interaction with others for hunting, defense, and reproduction and form small, long-lasting social groups called pods. In this species, pods do not have a highly organized social structure with a leader or dominant animal; instead, spinner pods are loose associations of a few key individuals, as well as dolphins who come and go. A pod of spinner dolphins may also spend time with other sea animals, such as pilot whales, spotted dolphins, or tuna.

Like most dolphins, spinners have good eyesight; however, they primarily rely on their sense of hearing and two kinds of voices, the sonic voice and the sonar voice, to let them know about their environment. The sonic or audible voice includes a vocabulary of clicks and whistles that are performed above or below the water. Along with these sounds, these mammals incorporate several mechanically produced sounds like jaw-snapping, flipper slapping, and crash dives. Sonic voice and mechanical sounds are associated with communication between animals. The sonar or echolocation voice is used to navigate. Spinners send out high frequency sounds that are reflected back to the senders as echoes. The dolphins listen for the echoes and use them to locate objects.

A female spinner calves (has offspring, called a calf) once every two or three years. After a gestation period of about 10 1/2 months, a single newborn, 29.5 to 33.5 inches (75 to 85 cm) long, is born. The calf is immediately pushed to the surface by its mother, so it can take in its first breath. A nursing mother lies on her side at the surface, enabling her offspring

Body Temperature

Animals that are described as warm blooded, or endothermic, maintain a constant internal temperature, even when exposed to extreme temperatures in their environment. In mammals, this internal temperature is about 97°F (36°C), while in birds, it is warmer, around 108°F (42°C).

Warm-blooded animals have developed several physiological and behavioral modifications that help regulate body temperature. Since their bodies generate heat by converting food into energy, they must take in enough food to fuel a constant body temperature. Once heat is produced, endotherms conserve it with insulating adaptations such as hair, feathers, or layers of fat. In extreme cold, they also shiver, a mechanism that generates additional heat.

Heart rate and rate of respiration in warm-blooded animals does not depend on the temperature of the surroundings. For this reason, they can be as active on a cold winter night as they are during a summer day. This is a real advantage that enables warm-blooded animals to actively look for food year round.

The internal temperature of cold-blooded, or ectothermic, animals is the same as the temperature of their surroundings. In other words, when it is hot outside, they are hot, and when it is cold outside, they are cold. In very hot environments the blood temperature of some cold-blooded animals can rise far above the blood temperature of warm-blooded organisms. Furthermore, their respiration rate is dependent on the temperature of their surroundings. To warm up and speed their metabolism, cold-blooded animals often bask in the sun. Therefore, cold-blooded animals such as fish, amphibians, and reptiles, tend to be much more active in warm environments than in cold conditions.

to feed and breathe. The mother's fat-rich milk supports the young dolphin for about seven months.

Spinner dolphins may use their pectoral fins to reach out and stroke each other, acts that strengthen the social bonds between them. Pairs of dolphins often swim along face to face, touching their flippers. Closely bonded animals, such as mother and calf, may swim in perfect synchrony as if mirror images of each other. The dolphins are also playful animals that make "toys" from materials in the environment and pass them back and forth to each other.

Humpback Whales

The humpback whale (*Megaptera novaeangliae*) is much bigger than the spinner dolphin, measuring 40 to 50 feet (12.9 to 15.2 m) long and weighing up to 55 tons. The common name *humpback* describes the motion this whale makes as it jumps out of the water. A typical humpback whale is black on the dorsal side, with a white ventral surface and distinctive 15-foot flippers on the sides of its body.

The head of a humpback whale is large in proportion to the rest of its torpedo-shaped body. Figure 6.3 shows that the mouth line runs high along the entire length of head, and the eyes are set above the ends of the mouth. The small ear slits are located behind and below the eyes.

On the top of its head, a humpback has a raised area in front of its two blowholes that functions like a splashguard to keep water from flowing back into the holes where it breathes. Rounded knobs called tubercles are also located on the head, often on the upper and lower jaws. Each tubercle

Fig. 6.3   Baleen whales have two blowholes and large mouths filled with baleen plates. (Courtesy Sanctuary Collection, NOAA)

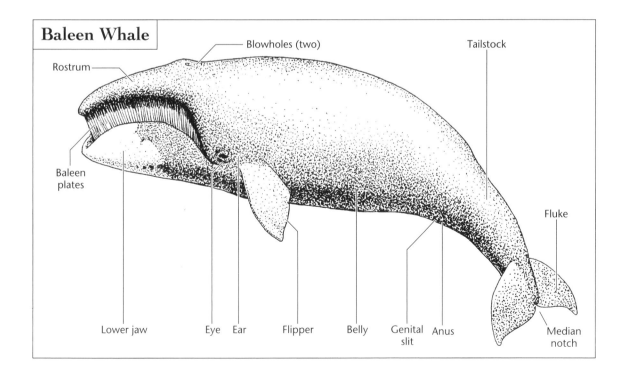

Baleen Whale

Blowholes (two)

Tailstock

Rostrum

Baleen plates

Lower jaw   Eye   Ear   Flipper   Belly   Genital   Anus
                                                               slit

Fluke

Median notch

contains a hairlike structure, a vibrissa, that is about 0.5 inch (1.3 cm) long. The vibrissa's function is not clearly understood, but it is believed to be important in detecting vibrations in the water. On the ventral side of the head, running from the tip of the lower jaw to the naval, there is an area of grooves known as ventral pleats, which are creased tissues that unfold when the whale opens its mouth, allowing the animal to expand the size of its bite to three times its normal width. The throat pleats can be seen when the whales breach, or jump in the water, as in the lower color insert on page C-8.

Whales are divided into two groups based on their feeding adaptations: the baleen whales and the toothed whales. Humpbacks are baleen whales, so named for the large plates in their mouths that act as food-catching sieves. Baleen is made of a flexible tissue that is chemically and physically similar to a fingernail. Plates are rooted in and grow from bases in the roof of the mouth. On each side of the upper jaw, there are 480 baleen plates. Each plate overlaps the adjacent one, forming dense mats that filter plankton from the water. The tongue wipes food off the plates and sweeps it into the whale's throat.

Several of the cetaceans migrate, and their paths and periods of migration vary by species. Migratory baleens divide their year between the rich feeding grounds of the cold seas and the warm oceans where they breed and calve.

Humpbacks live in small groups similar to the pods of spinner dolphins. From June to September, they feed in the waters around Alaska and in other cold regions where food is plentiful, leaving the area in early fall for the long trip to the tropics. Led by the sexually mature members of the group, the entire group makes the trip of about 3,500 miles (5,600 km) to warm coral reef waters, cruising at speeds of 2.3 miles per hour (2.0 kph), where females give birth.

Male humpback whales are extremely vocal and sing complex songs that can go on for hours. Within a population of whales, all of the males begin the breeding season singing the same song, but as the season progresses, each male creates his own version. By the end of a breeding season, individual songs have evolved so that every male's vocalizations are distinct. The exact functions of these songs are not known but are most likely associated with mating behaviors such as attracting females or warning off rival males.

Feeding occurs within the top 164 feet (50 m) of water. Humpbacks consume tons of plankton and krill, small, insectlike animals that live in the upper layers of water. To eat, a whale engulfs enormous gulps of water, then filters out food by sieving the water through the meshlike screen of baleen plates. Humpbacks have several feeding techniques, including one called bubble-netting. In this strategy, a whale dives beneath a school of prey and slowly begins to spiral upward around them, blowing bubbles as it goes. These bubbles herd the prey in the center of the circle. The whale then dives beneath the prey and swims up through the bubble net with its mouth open, gulping prey as it ascends.

Minke Whales

The minke whale (*Balaenoptera acutorostrata*) is the second smallest of the baleen whales, measuring about 32.8 feet (10 m) in length. A distinctive triangular head, narrow and pointed snout, and sickle-shaped dorsal fin make this whale easy to identify. Generally, minke whales are black, gray, or brown on their dorsal surface and a light color on their ventral surface. These active, agile whales have good maneuverability and speed, able to travel at 10.4 miles per hour (17 kph) for short periods of time.

As baleen whales, minkes are carnivores whose mouths are equipped with smooth baleen plates for filtering small organisms from gulps of water. Minkes can be found worldwide and are known to live in deep oceans, along coasts, and in coral reefs. Instead of seasonal migration, they only travel to follow their food. Sometimes minkes chase schools of small fish such as sardines and herring, swimming beneath and scooping them up in their open mouths. As in humpbacks, the throats of minkes are pleated so that they can expand their bite size.

Females enter their reproductive periods on 14-month cycles. Gestation lasts for 10 months, and calves are born in mid-winter. There is usually only one calf, but twins and triplets do occur. A newborn calf is only about 8.53 feet (2.6 m) long but grows quickly on its mother's milk, which supports the baby whale for four or five months. Young whales reach sexual maturity at seven years of age and live to be about 50 years old.

Minke whales are most often solitary animals, although they are also seen in small groups. When food is plentiful, several hundred animals may congregate in the feeding grounds where they communicate with grunts, clicks, and breaching. The sounds they produce are very low-frequency waves that can travel long distances under water. Worldwide populations of minke whales are larger than most other groups of whales, consisting of about a million animals. Because the whales are small, they have escaped predation by humans and maintained almost-normal population sizes.

Dugongs

Shy and retiring, dugongs are marine mammals that spend their days feeding in the shallow waters of reefs and along coastlines in the Indian Ocean, the Indonesian archipelago, and the southwestern Pacific around the Philippines. Even though they resemble a cross between a seal and a walrus, dugongs are more closely related to elephants. The slow-moving mammals are easily identified by their triangular, whalelike tails, broad trunklike snouts, and long bodies, which reach 9 feet (2.7 m). Dugongs have a thick layer of blubber under their skin, a feature that gives them a round-shouldered look. Their mouths look like vertical slits on their upper jaws, and their flippers are small and paddle shaped.

Dugongs have unusually slow metabolic rates for mammals but function well in their warm water environments where they float and feed, expending very little energy. With very few predators and plenty of food, migration and other energetic types of behavior are not necessary. Most of their time is spent grazing on sea grass blades and digging up the grass roots, their favorite parts of the plants. Roots of sea grasses are rich in carbohydrates, but to reach these treats, the animals must dig around on the bottom of the reef, behavior that has earned them the nickname "sea pigs." Equipped with very few teeth, a dugongs bites with a mobile disk at the end of its snout. The disk works like a rake, pulling in food and sending it back to the grinding plates in the mouth. The males also have tusks, enlarged incisors that project from below the upper lip.

Sexually mature between eight and 18 years, female dugongs give birth to one calf every three or four years. After a gestation period of 13 months, their cream-colored calves are born in shallow water. Mothers help the calves, measuring only 39.4 inches (100 cm), to the surface for their first breaths. A calf nurses for two years, always remaining close enough to its mother to touch her. During an average life span of 55 years, a female produces only five or six offspring.

Dugongs are so shy that not much is known about their social interactions. Attempts to observe their behavior disturb them and often kindle curiosity about the observers. Dugongs are spotted singly or in small groups of six to eight animals. Within a group there seems to be no leader or organized social structure.

Conclusion

Coral reef vertebrates include reptiles, birds, and mammals. Green sea turtles and hawksbill sea turtles are two reptiles that make their homes there. The green sea turtle, the largest of the hard-shelled turtle species, is a gentle vegetarian that can be found worldwide. Already extinct in some countries, its population is considered to be endangered. Populations of hawksbill sea turtles are even smaller, and this species has been on the endangered list for more than 40 years because its carapace is the source of "tortoise shell," which is still highly sought for jewelry and other accessories. Snake populations on reefs have fared better than those of turtles. Reef snakes are not aggressive, but they are venomous animals that many reef swimmers avoid.

The number and kinds of birds that live near coral reefs is quite large. Huge populations of terns, shearwaters, boobies, and frigate birds depend on the reef for food. Many are divers that plunge deep into the water to catch fish and squid. Others skim the water's surface, picking up zooplankton floating there. A lot of seabirds lay their eggs on nearby reef islands. Others are migratory animals that visit reefs on their way to breeding grounds.

Smaller in number, but larger in presence, are the marine mammals: the whales, dolphins, and dugongs. As with the birds, several species of mammals visit coral reefs, some staying for short periods. A few types of mammals spend the majority of their time there. The humpback and minke whales are regulars, along with the spinner dolphin. Around the reefs of Australia, docile dugongs live in small pods.

Vertebrates of coral reefs lean toward the spectacular, either physically large or brightly colored. The buoyancy of water enables these creatures to reach sizes that would never be possible on land. The use of color among vertebrates, as with other reef animals, sends clear, unambiguous messages about gender, danger, or age. As a group, vertebrates are obvious and essential members of the reef ecosystems.

Reefs in the Future

Few places are more highly praised for their beauty than coral reefs. These jeweled ecosystems, found off the coasts between the tropics of Cancer and Capricorn, can only survive in warm, tropical locales that offer the perfect combination of physical factors. Reefs form slowly, adding layers of skeletons to the reef structure over hundreds of years. In one year, a reef might boost its height by only 78.7 inches (200 cm). Estimates place the age of some of the older reefs at 10,000 to 30,000 years.

A mature coral reef provides habitats for more species of life than any other marine environment. All of the associated plants, invertebrates, and vertebrates live in close relationships with the coral animals. In the coral ecosystem, these communities thrive because nutrients and resources are tightly recycled, making them available to support many living things.

Corals, the small animals that build reefs, spend their lives inside limestone skeletons. Under ideal conditions, each coral can serve as host to thousands of algae cells. The algae are photosynthesizers that thrive in the clear reef water, making plenty of food for themselves as well as for the coral. To supplement their nutrition, corals also feed on small animals and organic matter in the water.

Coral reefs are not tolerant of changes in the physical conditions around them. Fluctuations in temperature, salinity, or the clarity of water can cause stress and damage. Since 1980, scientists have noticed that large sections of coral reefs have undergone bleaching. When this happens, the algae living in their tissues leave. If the stressors are short term, the algae often return to their hosts and the corals survive; however, if the stress is a serious long-lasting one, the coral animals die.

Human Impact

Stressors include a variety of events, some natural and some human made. Storms and changes in weather can alter the conditions of seawater around coral reefs, but most coral damage is the result of human activity. Exploitation of reefs, overfishing, increased rates of sedimentation in the water, and increased levels of nutrients in water are some of the most recent causes of coral death, but the chief problem appears to be global warming. Global warming, an increase in Earth's surface temperatures, is a direct result of burning coal and oil, two types of fuels made from the bodies of organisms that lived million of years ago.

Global warming raises temperatures in the ocean's waters. Since coral animals can only live within a narrow range of temperatures, every change is potentially fatal to them. Fluctuations in water temperature, either increases or decreases, can cause bleaching. Increased temperatures have the most dramatic effect. A rise in temperature of only 1.8–3.6°F (1–2°C) over several weeks can cause coral death.

In addition, the activities of people near reefs dramatically affect them. As communities on nearby islands and coasts increase in size, more homes, school, businesses, and industries are built. Construction loosens the soil and increases rates of erosion. As soil and sediment enter the clear, reef waters, the materials cloud the water, reducing the light that can reach the corals and their single-celled algae. The lifestyle of humans also raises the levels of nutrients that enter seawater. Nutrients increase the growth rate of algae in water, causing further shading of delicate corals.

A Change in Thinking

As public awareness increases, concern for the safely of coral reefs grows. Marine scientists are calling for an immediate reduction in the levels of greenhouse gas emissions and marine pollution. In addition, reserves of protected marine areas are being established. Some reserves contain artificial reefs made of cement blocks or old tires, and these structures seem to help replenish populations of reef fish.

Both national and international efforts have been launched to protect these delicate ecosystems. The Global Coral Reef Monitoring Network (GCRMN), staffed by the United Nations and several national governments, keeps an eye on the condition of reefs worldwide. The goal of this organization is to provide information for reef research and to raise awareness of the problems and solutions concerning reefs. The Coral Reef Task Force is monitoring and mapping reefs off U.S. coasts. All of this attention seems to be making a difference. There is evidence that reefs are already improving in health in areas where they are protected from pollution, sedimentation, and overuse.

Glossary

A

algal bloom   The rapid growth of cyanobacteria or algae populations that results in large mats of organisms floating in the water.

amphibian   A cold-blooded, soft-skinned vertebrate whose eggs hatch into larvae that metamorphose into adults.

animal   An organism capable of voluntary movement that consumes food rather than manufacturing it from carbon compounds.

anterior   The region of the body that is related to the front or head end of an organism.

appendage   A structure that grows from the body of an organism, such as a leg or antenna.

arthropod   An invertebrate animal that has a segmented body, joined appendages, and chitinous exoskeleton.

asexual reproduction   A type of reproduction that employs means other than the union of an egg and sperm. Budding and binary fission are forms of asexual reproduction.

autotroph   An organism that can capture energy to manufacture its own food from raw materials.

B

binary fission   A type of cell division in monerans in which the parent cell separates into two identical daughter cells.

biodiversity   The number and variety of life-forms that exist in a given area.

bird   A warm-blooded vertebrate that is covered with feathers and reproduces by laying eggs.

bladder   In macroalgae, an inflatable structure that holds gases and helps keep blades of the plant afloat.

blade   The part of a nonvascular plant that is flattened and leaflike.

brood   A type of behavior that enables a parent to protect eggs or offspring as they develop.

budding   A type of asexual reproduction in which an offspring grows as a protrusion from the parent.

buoyancy   The upward force exerted by a fluid on matter that causes the matter to tend to float.

C

carnivore   An animal that feeds on the flesh of other animals.

chanocyte   A flagellated cell found in the gastrovascular cavity of a sponge that moves water through the pores, into the gastrovascular cavity, and out the osculum (an exit for outflow).

chitin   A tough, flexible material that forms the exoskeletons of arthropods and cell walls of fungi.

chlorophyll   A green pigment, found in all photosynthetic organisms, that is able to capture the Sun's energy.

cilia   A microscopic, hairlike cellular extension that can move rhythmically and may function in locomotion or in sweeping food particles toward an animal's mouth or oral opening.

cnidarian   An invertebrate animal that is radially symmetrical and has a saclike internal body cavity and stinging cells.

cnidocyte   A nematocyst-containing cell found in the tentacles of cnidarians that is used to immobilize prey or defend against predators.

countershading   One type of protective, two-tone coloration in animals in which surfaces that are exposed to light are dark colored and those that are shaded are light colored.

cyanobacteria   A moneran that contains chlorophyll as well as other accessory pigments and can carry out photosynthesis.

D

detritivore   An organism that feeds on dead and decaying matter.

detritus   Decaying organic matter that serves as a source of energy for detritivores.

DNA   Deoxyribonucleic acid, a molecule located in the nucleus of a cell that carries the genetic information that is responsible for running that cell.

dorsal   Situated on the back or upper side of an organism.

E

ecosystem   A group of organisms and the environment in which they live.

endoskeleton   An internal skeleton or support system such as the type found in vertebrates.

energy   The ability to do work.

epidermis   The outer, protective layer of cells on an organism, such as the skin.

exoskeleton   In crustaceans, a hard but flexible outer covering that supports and protects the body.

F

fish   A cold-blooded, aquatic vertebrate that has fins, gills, and scales and reproduces by laying eggs that are externally fertilized.

flagellum   A long, whiplike cellular extension that is used for locomotion or to create currents of water within the body of an organism.

food chain   The path that nutrients and energy follow as they are transferred through an ecosystem.

food web   Several interrelated food chains in an ecosystem.

fungus   An immobile heterotrophic organism that consumes its food by first secreting digesting enzymes on it, then absorbing the digested food molecules through the cell walls of threadlike hyphae.

G

gastrodermis   The layer of cells that lines the digestive cavity of a sponge or cnidarian, and the site at which nutrient molecules are absorbed.

gastropod   A class of arthropods that has either one shell or no shells, a distinct head equipped with sensory organs, and a muscular foot.

gill   A structure containing thin, highly folded tissues that are rich in blood vessels and serve as the sites where gases are exchanged in aquatic organisms.

glucose   A simple sugar that serves as the primary fuel in the cells of most organisms. Glucose is the product of photosynthesis.

H

herbivore   An animal that feeds on plants.

hermaphrodite   An animal in which both male and female sexual organs are present.

heterotroph   An organism that cannot make its own food and must consume plant or animal matter to meet its body's energy needs.

holdfast   The rootlike portion of a macroalga that holds the plant to the substrate.

hydrogen bond   A weak bond between the positive end of one polar molecule and the negative end of another.

hyphae   Filamentous strands that make up the bodies of fungi and form the threadlike extensions that produce digestive enzymes and absorb dissolved organic matter.

I

invertebrate   An animal that lacks a backbone, such as a sponge, cnidarian, worm, mollusk, or arthropod.

L

lateral   The region of the body that is along the side of an organism.

lateral line   A line along the side of a fish that connects to pressure-sensitive nerves that enable the fish to detect vibrations in the water.

larva   The newly hatched offspring of an animal that is structurally different from the adult form.

light   A form of electromagnetic radiation that includes infrared, visible, ultraviolet, and X-ray that travels in waves at the speed of 186,281 miles (300,000 km) per second.

M

mammal   A warm-blooded vertebrate that produces living young that are fed with milk from the mother's mammary glands.

mantle   A thin tissue that lies over the organs of a gastropod and secretes the shell.

mesoglea   A jellylike layer that separates the two cell layers in the bodies of sponges and cnidarians.

milt   A fluid produced by male fish that contains sperm and is deposited over eggs laid by the female.

mixotroph   An organism that can use the Sun's energy to make its own food or can consume food.

molt   Periodic shedding of an outer layer of shell, feathers, or hair that allows new growth to occur.

moneran   A simple, one-celled organism that neither contains a nucleus nor membrane-bound cell structures.

motile   Capable of moving from place to place.

N

nematocyst   In cnidarians, a stinging organelle that contains a long filament attached to a barbed tip that can be used in defense or to capture prey.

O

omnivore   An animal that eats both plants and animals.

operculum   In fish, the external covering that protects the gills. In invertebrates, a flap of tissue that can be used to close the opening in a shell, keeping the animal moist and protecting it from predators.

oviparous   An animal that produces eggs that develop and hatch outside the mother's body.

ovoviviparous   An animal that produces eggs that develop and hatch within the mother's body, then are extruded.

P

pectoral An anatomical feature, such as a fin, that is located on the chest.

pelvic An anatomical feature, such as a fin, that is located near the pelvis.

photosynthesis The process in which green plants use the energy of sunlight to make nutrients.

plant A nonmotile, multicellular organism that contains chlorophyll and is capable of making its own food.

polar molecule A molecule that has a negatively charged end and a positively charged end.

polychaete A member of a group of worms that has a segmented body and paired appendages.

posterior The region near the tail or hind end of an organism.

productivity The rate at which energy is used to convert carbon dioxide and other raw materials into glucose.

protist A one-celled organism that contains a nucleus and membrane-bound cell structures such as ribosomes for converting food to energy and Golgi apparati for packaging cell products.

R

radula A long muscle used for feeding that is covered with toothlike projections, found in most types of gastropods.

reptile A cold-blooded, egg-laying terrestrial vertebrate whose body is covered with scales.

S

salinity The amount of dissolved minerals in ocean water.

school A group of aquatic animals swimming together for protection or to locate food.

sessile Permanently attached to a substrate and therefore immobile.

setae Hairlike bristles that are located on the segments of polychaete worms.

sexual reproduction A type of reproduction in which egg and sperm combine to produce a zygote.

spawn The act of producing gametes, or offspring, in large numbers, often in bodies of water.

spicule In sponges, a needle-like, calcified structure located in the body wall that provides support and protection.

spiracle An opening for breathing, such as the blowhole in a whale or the opening on the head of a shark or ray.

stipe A stemlike structure in a nonvascular plant.

surface tension   A measure of how easy or difficult it is for molecules of a liquid to stick together due to the attractive forces between them.

swim bladder   A gas-filled organ that helps a fish control its position in the water.

symbiosis   A long-term association between two different kinds of organisms that usually benefits both in some way.

T

territorial behavior   The defense of a certain area or territory by an animal for the purpose of protecting food, a mate, or offspring.

thallus   The body of a macroalgae, made up of the blade, stipe, and holdfast.

V

ventral   Situated on the stomach or lower side of an organism.

vertebrate   A member of a group of animals with backbones, including fish, amphibian, reptiles, birds, and mammals.

viviparous   An animal that gives birth to living offspring.

Z

zooxanthella   A one-celled organism that lives in the tissues of invertebrates such as coral, sponge, or anemone where it carries out photosynthesis.

Further Reading and Web Sites

Books

Banister, Keith, and Andrew Campbell. *The Encyclopedia of Aquatic Life.* New York: Facts On File, 1985. Well written and beautifully illustrated book on all aspects of the ocean and the organisms in it.

Coulombe, Deborah A. *The Seaside Naturalist.* New York: Fireside, 1990. A delightful book for young students who are beginning their study of ocean life.

Davis, Richard A. *Oceanography: An Introduction to the Marine Environment.* Dubuque, Iowa: Wm. C. Brown Publishers, 1991. A text that helps students become familiar with and appreciate the world's oceans.

Dean, Cornelia. *Against the Tide.* New York: Columbia University Press, 1999. An analysis of the impact of humans and nature on the ever-changing beaches.

Ellis, Richard. *Encyclopedia of the Sea.* New York: Alfred A. Knopf, 2000. A factual, yet entertaining, compendium of sea life and lore.

Garrison, Tom. *Oceanography.* New York: Wadsworth Publishing, 1996. An interdisciplinary examination of the ocean for beginning marine science students.

Karleskint, George, Jr. *Introduction to Marine Biology.* Belmont, Calif.: Brooks/Cole-Thompson Learning, 1998. An enjoyable text on marine organisms and their relationships with one another and with their physical environments.

McCutcheon, Scott, and Bobbi McCutcheon. *The Facts On File Marine Science Handbook.* New York: Facts On File, 2003. An excellent resource that includes information on marine physical factors and living things as well as the people who have been important in ocean studies.

Nowak, Ronald M., et al. *Walker's Marine Mammals of the World.* Baltimore, Md.: Johns Hopkins University Press, 2003. An overview on the anatomy, taxonomy, and natural history of the marine mammals.

Pinet, Paul R. *Invitation to Oceanography.* Sudbury, Mass.: Jones and Bartlett Publishers, 2000. Includes explanations of the causes and effects of tides and currents, as well as the origins of ocean habitats.

Prager, Ellen J. *The Sea.* New York: McGraw-Hill, 2000. An evolutionary view of life in the Earth's oceans.

Reeves, Randall R., et al. *Guide to Marine Mammals of the World.* New York: Alfred A. Knopf, 2002. An encyclopedic work on sea mammals accompanied with gorgeous color plates.

Rice, Tony. *Deep Oceans*. Washington, D.C.: Smithsonian Museum Press, 2000. A visually stunning look at life in the deep ocean.

Sverdrup, Keith A., Alyn C. Duxbury, and Alison B. Duxbury. *An Introduction to the World's Oceans*. New York: McGraw Hill, 2003. A comprehensive text on all aspects of the physical ocean, including the seafloor and the ocean's physical properties.

Thomas, David. *Seaweeds*. Washington, D.C.: Smithsonian Museum Press, 2002. Illustrates and describes seaweeds from microscopic forms to giant kelps, explaining how they live, what they look like, and why humans value them.

Thorne-Miller, Boyce, and John G. Catena. *The Living Ocean*. Washington, D.C.: Friends of the Earth, 1991. A study of the loss of diversity in ocean habitats.

Waller, Geoffrey. *SeaLife: A Complete Guide to the Marine Environment*. Washington, D.C.: Smithsonian Institution Press, 1996. A text that describes the astonishing diversity of organisms in the sea.

Web Sites

Bird, Jonathon. *Adaptations for Survival in the Sea,* Oceanic Research Group, 1996. Available online. URL: http://www.oceanicresearch.org/adapspt.html. Accessed March 19, 2004. A summary and review of the educational film of the same name, which describes and illustrates some of the adaptations that animals have for life in salt water.

Buchheim, Jason. "A Quick Course in Ichthyology." Odyssey Expeditions. Available online. URL: http://www.marinebiology.org/fish.htm. Accessed January 4, 2004. A detailed explanation of fish physiology.

"Conservation: Why Care About Reefs?" REN Reef Education Network, Environment Australia. Available online. URL: http://www.reef.edu.au/asp_pages/search.asp. Accessed November 18, 2004. A superb Web site dedicated to the organisms living in and the health of the coral reefs.

Duffy, J. Emmett. "Underwater urbanites: Sponge-dwelling napping shrimps are the only known marine animals to live in colonies that resemble the societies of bees and wasps." *Natural History*. December 2003. Available online. URL: http://www.findarticles.com/cf_dls/m1134/10_111736243/print.jhtml. Accessed January 2, 2004. A readable and fascinating explanation of eusocial behavior in shrimp and other animals.

"Fungus Farming in a Snail." *Proceedings of the National Academy of Science,* 100, no. 26 (December 4, 2003). Available online. URL: http://www.pnas.org/cgi/content/abstract/100/26/15643. A well-written, in-depth analysis of the ways that snails encourage the growth of fungi for their own food.

Gulf of Maine Research Institute Web site. Available online. URL: http://www.gma.org/about_GMA/default.asp. Accessed January 2, 2004. A comprehensive and up-to-date research site on all forms of marine life.

"Habitat Guides: Beaches and Shorelines." eNature. Available online. URL: http://www.enature.com/habitats/show_sublifezone.asp?sublifezoneID=60#Anchor-habitat-49575. Accessed November 21, 2003. A Web site with young people in mind that provides comprehensive information on habitats, organisms, and physical ocean factors.

Huber, Brian T. "Climate Change Records from the Oceans: Fossil Foraminifera." Smithsonian National Museum of Natural History. June 1993. Available online. URL: http://www.nmnh.si.edu/paleo/marine/foraminifera.htm. Accessed December 30, 2003. A concise look at the natural history of foraminifera.

"Index of Factsheets." Defenders of Wildlife. Available online. URL: http://www.kidsplanet.org/factsheets. Accessed November 18, 2004. Various species of marine animals are described on this excellent Web site suitable for both children and young adults.

King County's Marine Waters Web site. Available online. URL: http://splash.metrokc.gov/wlr/waterres/marine/index.htm. Accessed December 2, 2003. A terrific Web site on all aspects of the ocean, emphasizing the organisms that live there.

Mapes, Jennifer. "U.N. Scientists Warn of Catastrophic Climate Changes." National Geographic News. February 6, 2001. Available online. URL: http://news.nationalgeographic.com/news/2001/02/0206_climate1.html. A first-rate overview of the current data and consequences of global warming.

National Oceanic and Atmospheric Administration Web site. Available online. URL: http://www.noaa.gov/. A top-notch resource for news, research, diagrams, and photographs relating to the oceans, coasts, weather, climate, and research.

"Resource Guide, Elementary and Middle School Resources: Physical Parameters." Consortium for Oceanographic Activities for Students and Teachers. Available online. URL: http://www.coast-nopp.org/toc.html. Accessed December 10, 2003. A Web site for students and teachers that includes information and activities.

"Sea Snakes in Australian Waters." CRC Reef Research Centre. Available online. URL: http://www.reef.crc.org.au/discover/plantsanimals/seasnakes. Accessed November 18, 2004. An overview of sea snake classification, breeding, and venom.

U.S. Fish and Wildlife Service Web site. Available online. URL: http://www.fws.gov/. A federal conservation organization that covers a wide range of topics, including fisheries, endangered animals, the condition of the oceans, and conservation news.

Index